Scientific Simulations with Special-Purpose Computers — the GRAPE systems

Scientific Simulations with Special-Purpose Computers — the GRAPE systems

Junichiro Makino
University of Tokyo, Japan

Makoto Taiji
Institute for Statistical Mathematics, Tokyo, Japan

JOHN WILEY & SONS
Chichester · New York · Weinheim · Brisbane · Singapore · Toronto

Other Wiley Editorial Offices

John Wiley & Sons, Inc., 605 Third Avenue,
New York, NY 10158-0012, USA

WILEY-VCH Verlag GmbH, Pappelallee 3,
D-69469 Weinheim, Germany

Jacaranda Wiley Ltd, 33 Park Road, Milton,
Queensland 4064, Australia

John Wiley & Sons (Asia) Pte Ltd, Clementi Loop #02-01,
Jin Xing Distripark, Singapore 129809

John Wiley & Sons (Canada) Ltd, 22 Worcester Road,
Rexdale, Ontario M9W 1L1, Canada

Library of Congress Cataloging-in-Publication Data

Makino, Junichiro
    Scientific simulations with special purpose computers : The GRAPE
Systems / Junichiro Makino, Makoto Taiji.
      p.   cm.
    Includes bibliographical references and index.
    ISBN 0 471 96946 X (cloth : alk. paper)
    1. Science — Computer simulation. 2. GRAPE (Computer file)
I. Taiji, Makoto. II. Title
Q183.9.M35   1998   97–44584
521—dc21

British Library Cataloguing in Publication Data

A catalogue record for this book is available from the British Library

ISBN 0 471 96946 X

Typeset by the author
Printed and bound in Great Britain by Bookcraft (Bath) Ltd
This book is printed on acid-free paper responsibly manufactured from sustainable forestry,
in which at least two trees are planted for each one used for paper production

Contents

1

Introduction

The goal of this book is to describe a relatively new and unexplored way of doing computational science, namely building computers, or "computing devices", specialized to solve particular kind of problems. In this book, we call such computing devices "special-purpose computers", opposed to general-purpose computers.

A general-purpose programmable computer can, in principle, solve *any* numerical problem, if it can spend an infinite amount of time and memory. Thus, we build special-purpose computers because they can solve a problem faster than a general-purpose computer of the same price. In other words, the only advantage of a special-purpose computer is that it offers a better cost-performance than a general-purpose computer.

There have been a number of projects to develop computers specialized to limited problems. However, not all projects have been successful. In fact, only a handful of projects can be regarded as having been highly successful. In this book, we describe the GRAPE project, which, we believe, has been highly successful, and analyze several other projects to understand why one project was a success while another was not. Through this analysis, we hope to present some sort of "general theory of special-purpose computers".

This book is organized in the following way. In Chapter 2, we present a brief overview of the history of general-purpose computers. We focus on the relation between the evolution of semiconductor technology and that of the architecture and performance of general-purpose systems. To take advantage of the evolution of semiconductor technology, the architecture of general-purpose computers has become more and more complex. Thus, it becomes increasingly difficult to build high-performance general-purpose computers. Even so, the improvement in the performance of the general-purpose computers has been significantly slower than it could be if the advances in semiconductor technology had been fully utilized. We describe the reason for this discrepancy. In Chapter 3, we examine the special-purpose computers for computational sciences. All of them tried to achieve better performance than general-purpose computers by using semiconductor technology in more streamlined ways. We analyze the basic concepts and actual implementations. It is shown that the special-purpose systems took advantage of advance in semiconductor technology which were not fully explored by contemporary general-purpose systems. It is also demonstrated that much of the technological advantage vanished as semiconductor technology evolved further.

In Chapters 4 through 6, the GRAPE systems, our approach to building and using special-purpose computers, are described. We first overview the history of the GRAPE project in Chapter 4, and also present some technical details of the hardware. In Chapter 5, we describe the numerical algorithms and their implementation on GRAPE hardware. Finally, in Chapter 6 we overview the scientific results obtained using GRAPE hardware.

In Chapter 7, we speculate on the future of special-purpose computers. We present an example architecture of the GRAPE system in the near future, and discuss several design trade-offs.

2

The Evolution of General-purpose Computers

The evolution of the general-purpose computer in the last half century is quite amazing. If we compare the speed of the present fastest parallel computer (as of early 1997, the Intel-Sandia ASCI Red machine, with a peak speed exceeding 1 Tflops, is currently the fastest general-purpose computer) to that of, say, the ENIAC, the ASCI Red is almost 10^9 times faster than the ENIAC. Roughly speaking, the speed of the fastest computer has been increasing by a factor of 100 in every 10 years, and this trend is expected to continue for at least the next 10–20 years.

In the following, we first analyze how this impressive growth in speed was achieved. Then, we consider the possibilities of doing even better.

The speed of a computer for a problem is expressed as

$$R = fnc^{-1}\eta, \tag{2.1}$$

where R is the computing speed (in terms of the number of floating point operations per second, for example) and f, n, c and η are the clock frequency, the number of floating point operation units, the average number of clock cycles taken to perform one floating point operation, and the efficiency of the computer achieved on the problem, respectively.

For example, if a machine with $f = 100$MHz, $n = c = 1$ achieved 50% efficiency for a problem, the speed achieved is 50 Mflops. For a machine with $f = 10$MHz, $n = 1000$, $c = 10$ and 10% efficiency, the speed achieved is 100 Mflops.

Roughly speaking, f is determined by the device technology. If the logic design is the same, the clock frequency is determined ultimately by the switching speed of the transistor used. The other two hardware parameters, n and c, are to some extent complementary. For a given number of transistors, the designer can either design a machine with many processors, which take many cycles to perform one floating point operation, or alternatively, he can design a machine with a relatively small number of processors which take a small number

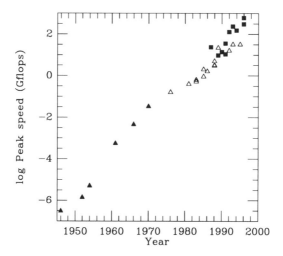

Figure 1   The peak speed of representative computers. Filled triangles are the
speed of non-vector "supercomputers" (before Cray-1). Open triangles are the
shared-memory vector machines, and filled squares are parallel machines.

of cycles to perform one operation. If the efficiency is the same, the speed
achieved would not be much different. Thus, $n_{ops} = n/c$, the average number
of floating point operations performed per clock cycle, is a more appropriate
measure of both the speed and physical size of the machine.

For the time being, we assume that the efficiency is somehow maintained to
be not too far from 100%. Figures 2 and 3 gives the clock frequency f and the
maximum number of floating point operations per clock n_{ops} for the machines
shown in Figure 1. If we look at Figure 3, it is clear that the machines called
"MPPs" (massively-parallel processors) and vectors are well separated from
each other, but the rate of evolution is actually similar. Roughly speaking, n_{ops}
of vector machines increased by a factor of 10 in 10 years. The rate of increase
of n_{ops} for MPPs is more difficult to determine because of a much larger
scatter, but clearly it is not faster than the evolution rate of vector machines.
In 1987, Thinking Machines Corporation was able to deliver a machine with
4096 fully-pipelined floating point units. No companies in 1997 have shipped
a machine with more than 10 000 floating point units.

When we look at Figure 2, however, the difference between vector machines
and MPPs is drastic. The clock speed of vector machines was improved by
roughly a factor of 10 in the last 20 years, while that of MPPs was improved
by nearly a factor of 100 in the last 10 years.

This very rapid evolution was lead by advances in microprocessor technology, which in turn was lead by advance in CMOS VLSI technology. Figure

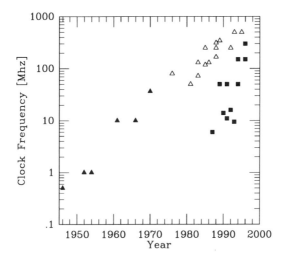

Figure 2 The clock frequency of representative computers. Filled triangles are the speed of non-vector "supercomputers" (before Cray-1). Open triangles are the shared-memory vector machines, and filled squares are parallel machines.

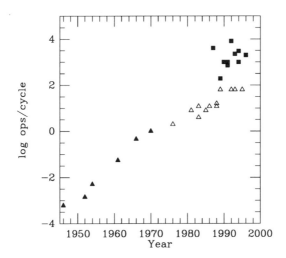

Figure 3 As Figure 2 but showing the number of floating point operations performed per clock cycle.

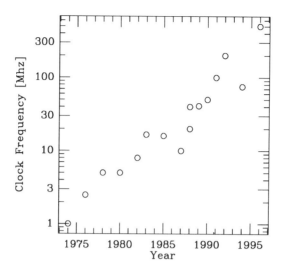

Figure 4 The clock frequency of representative microprocessors.

4 shows the clock speed of representative one-chip microprocessors plotted versus the time of the first delivery. Roughly speaking, an improvement of a factor of 100 was achieved in the last 15 years. The clock speed of MPPs is essentially the same as (but somewhat slower than) the clock speed of microprocessors of the same period, since most MPPs are actually either based on microprocessors or on essentially the same technology. Because the market for MPPs is much smaller than that for PCs or workstations, manufacturers of MPP systems are forced to use somewhat outdated technology and/or somewhat less optimized design to keep the development cost within an acceptable range.

The vector machines shown in Figures 1–3 are based on different device technologies from CMOS (typically ECL). The difference between ECL and CMOS is that ECL is faster than CMOS if the manufacturing technology is the same, but at the same time, ECL dissipates much more heat.

Until around 1990, the difference in clock speed between ECL-based vector machines and that of CMOS microprocessors had been large. However, in the 1990s the differences have become very small. In fact, the announcement of a 600MHz DEC Alpha 21164 chip in early 1997 means that the fastest CMOS microprocessor has a clock frequency higher than that of any other computer. This is essentially because the switching time of transistors no longer makes up the largest fraction of the total signal delay. The time taken for a signal to travel from one place to another (transmission delay) has become as important as the switching speed of the transistor itself (gate delay). With

Table 1 Memory bandwidth relative to the computing speed.

Machines	word/flops
DEC 8400 5/300(6 procs)	0.036
SGI PowerChallenge (8 procs, 75 MHz R8000)	0.039
Sun Ultra-1	0.093
HP J-210	0.095
IBM RS/6000 990	0.31
Cray T-90 (1 proc)	0.95
NEC SX-3-12	0.88

deep-submicron technology becoming available in the mid- and late-90s, the transmission delay will become the main part of the total signal delay.

CMOS technology has an important advantage over ECL technology in that the heat dissipation is smaller. As a result, one can pack a much larger number of transistors in the same space, and therefore the length of the wires between transistors is smaller for CMOS-based processors. Thus, it is quite natural that the clock speed of CMOS microprocessor has become roughly the same as that of vector machines.

Of course, so far we have ignored one important factor: the memory bandwidth. The clock speed of a single-chip microprocessor has been improved enormously in the last 20 years, but the data transfer rate between the memory and processor was not improved at a similar rate. Thus, the performance of many real applications is quite noticeably lower on workstations or MPPs compared to that on vector machines. This problem of bandwidth is actually more deep-rooted than is sometimes stated.

The apparent memory bandwidth problem is that the data transfer rate of the main memory of present workstations and Shared-Memory Multiprocessors (SMPs) is too small compared to the speed of the processors themselves. Table 1 gives the number of words the processor can transfer between the main memory and the CPU in the time it performs one floating point operation, in units of the number of 64-bit words. This table is based on the result of McCulpin's STREAM benchmark [McC95].

Vector machines (Cray T-90 and NEC SX-3) maintain data transfer rates of around one word/flop, while SMPs (DEC 8400 and SGI PowerChallenge) and high-end workstations (Sun Ultra-I, HP model J-210) fall below 0.1 (IBM RS/6000 achieves a much better rate, though).

The problem of limited main-memory bandwidth, in its present form, comes mainly from the design decision which traditionally almost ignored the necessity of the main memory bandwidth, and hopefully will be much improved in the next few years. For example, one of the most influential textbooks on computer architecture of the early 1990s [HP90] barely discusses main memory bandwidth. In fact, the framework of their "quantitative" analysis ignores the importance of memory bandwidth. Fortunately, in recent years computer architects have begun to realize that memory bandwidth is very important in achieving balanced performance for real applications.

If we calculated the cost of workstations or SMPs per main memory bandwidth, they are by far more expensive than low-end vector machines such as the Cray J-90, Fujituu VPP-300 and NEC SX-4. There is no technological/economical reason why the main memories of present workstations and SMPs cannot be much faster.

The limitation in the bandwidth between the CPU and whatever is connected to it (usually an off-chip cache memory) is actually more severe than the current limitation in the bandwidth of the main memory. The CPU chip is connected to the off-chip cache by external wirings through pins. The number of pins one can use for one chip has not grown very rapidly. Moreover, the cycle time of the data transfer through the I/O pin tends to be much longer than the clock frequency of the CPU chip itself. Both are quite severe limitations.

The pin count is rather difficult to increase. Early microprocessors in the late 1970s typically had 40 pins, while recent microprocessors have around 500. The data transfer rate through an I/O pin is difficult to improve, essentially because the connection is physically long. In principle, one can drive a long line with a high clock speed by using a matched transmission line. However, this technique greatly increases the power dissipation.

This limitation in the data transfer bandwidth between CPU and off-chip memory is at present not so pronounced because it is caught up with yet another problem, namely the limitation in the degree of parallelism.

The number of transistors one can integrate on a single chip has been increasing at the constant rate of a factor of four in three years. In 1996, an LSI chip can integrate more than 10 million transistors. This number is very large compared to the number of transistors needed to implement a fully pipelined floating-point unit. It requires only 10^5 transistors to implement a double-precision floating-point multiplier. Thus, in theory, one LSI chip can pack in about 100 floating-point units. However, as of 1997, no microprocessor has more than four floating-point units (two adders and two multipliers), and most of them have two (one adder and one multiplier).

Figure 5 plots n_{ops}, the number of floating point operations that can be performed in single cycle, for various microprocessors. In the 1980s the growth of n_{ops} was very rapid. Intel 8087, announced in 1980, took nearly 100 cycles to perform one operation. Intel i860, announced in 1988, can perform 1.5 floating-point operations per cycle. However, since 1988, there has been very little advance in n_{ops}.

The reason why n_{ops} did not go beyond 2 is essentially the same as the reason why recent "superscalar" microprocessors still issue typically only 4–6 instructions, and on average execute less than two instructions. A decent

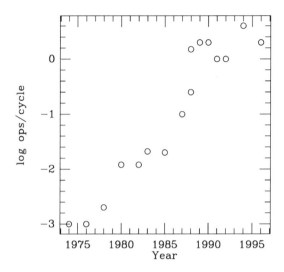

Figure 5 The number of floating point operations per cycle of representative microprocessors.

32-bit single-issue RISC CPU can be constructed with much less than 10^5 transistors. Thus, a single VLSI of today should be able to integrate about 100 such CPUs, resulting in an increase of about a factor of 50 compared to the speed of a 4-issue superscalar CPU. Why has such a chip not been developed?

The reason is that the software cannot utilize too many functional units, or at least that the designers of present microprocessors believe so. As long as the new processor has to run existing benchmark programs (for example, the SPECfp92/95 suites), they are perfectly right. Many detailed simulations have proved that it is very difficult for a compiler to generate the assembly code which can keep more than four functional units busy for a wide range of programs. Thus, if the goal of designing a microprocessor is to run a wide range of existing applications, there is no reason to go further than four functional units.

If we give up on a balanced performance for a wide range of applications, it is not very difficult to use a number of processors simultaneously. In other words, for some problems, state-of-the-art microprocessors do not make best use of the vast number of transistors available on a single LSI chip. However, the limitation in the off-chip bandwidth actually makes it impractical to manufacture a chip with more than 2–4 floating-point units.

The challenge that designers of present-day microprocessors are faced with is similar to that designers of high-end computers encountered in the 1970s.

Table 2 The SIA prediction for the memory bandwidth and the computing speed of a chip.

Year	1995	1998	2001	2004	2007	2010
Design rule (μm)	0.35	0.25	0.18	0.13	0.10	0.07
I/O Bus width (bit)	64	64	128	128	256	256
I/O clock (MHz)	50	66	100	100	125	150
I/O Bandwidth (GB/s)	0.4	0.53	1.6	1.6	4	4.8
Transistor Count	12M	28M	64M	150M	350M	800M
Chip clock (MHZ)	150	200	300	400	500	625
Chip performance (Gflops)	1.8	5.6	19.2	60	175	500
Balance (word/flops)	0.027	0.012	0.104	0.0033	0.0028	0.0012

The number of transistors available had exceeded the number necessary to implement a fully-pipelined CPU which can execute one instruction per cycle. The designers of high-end machines tried to make use of the large number of transistors available on the system by using multiple pipelines per processor and multiple processors. However, it is not easy to repeat this solution with a microprocessor, since one has to increase the memory bandwidth, i.e., chip-to-chip communication bandwidth.

Table 2 gives a prediction of the available off-chip bandwidth and the speed of floating-point operation of a single VLSI, based on the roadmap of silicon VLSI technology made by SIA [SIA94]. Here we assumed that 10% of the total of the transistors is used for the floating-point units, which is somewhat larger than what is typically used in present-day microprocessors, but was quite usual several years ago. We can see that the discrepancy between the bandwidth and the potential performance of floating-point operations grows quite rapidly. In 10 years, the bandwidth relative to the calculation speed will be reduced by a factor of 10, even though the absolute bandwidth will be improved by a factor of 10.

Whether this prediction is accurate or not remains to be seen, partly because the SIA projection is rather conservative in extrapolating the advances made in packaging and interconnection. However, even if the present ratio between the calculation speed and the external bandwidth is maintained by some technological breakthrough, the off-chip bandwidth has already fallen far short of the computing speed.

To summarize this chapter, the history of the general-purpose computer is the history of the improvement of the clock speed and the number of operations per clock period. The improvement in the speed of ECL-based systems reached a plateau, and CMOS-based systems, though the rate of improvement is slowing down, will become faster than ECL-based systems.

The number of floating-point operations per clock cycle of high-end computers improved fairly quickly until it reached two operations per clock with the CDC 7600. After that, the improvement slowed down. Vector machines of the 1990s typically have only 32–64 arithmetic pipelines, even though the number of transistors in these machines is about 1000, as large as that in CDC 7600. This is mainly because of the difficulties in constructing a memory system that has sufficient bandwidth to keep all the arithmetic pipelines busy.

In the 1980s, MPPs seemed to be the breakthrough. However, if we look at the evolution of MPPs themselves, we find that their evolution had also slowed down by 1990. The reason is exactly the same as the reason why vector machines reached a plateau. It's not easy to increase the bandwidth between the memory and processor. The growth in the number of floating-point units in one microprocessor practically stopped after the Intel i860 achieved 1.5 operations per clock.

In other words, the CMOS single-chip microprocessors now have exactly the same problem as the ECL-based high-end computers had in the mid-1970s. ECL-based computers did not really solve the problem, in the sense that their evolution slowed down in the mid-1980s. The performance plateau for CMOS microprocessors also seems to have begun.

Note that even with this limitation, the performance of high-performance computers has improved by a factor of 100 in 10 years, and will keep doing so. What we have seen so far is that this rate of increase is actually much lower than what could have been achieved. Roughly speaking, the number of transistors on a single VLSI chip quadruples every three years, and the clock speed of the chip also doubles in the same three years. Thus, if we can keep using the same percentage of the total number of transistors to actually perform arithmetic operations, the total processing speed of a computer will increase by a factor of eight in three years, which is a factor of 1000 in 10 years. Thus, on average, general-purpose computers have been becoming less and less efficient, at the rate of a factor of 10 in 10 years, for at least last two decades.

If we could utilize a larger percentage of the silicon for doing floating-point operations, we could have achieved growth of a factor of 1000 in 10 years. How we can achieve this faster growth by giving up the generality will be addressed in the next two chapters.

3

Overview of Special-Purpose Systems

3.1 The trouble with the von-Neumann architecture

In the previous chapter, we have seen that the slowing down in the growth of the performance of both ECL-based computers and CMOS-based microprocessors has a single root cause. It is difficult to utilize more than a handful of arithmetic units. This difficulty stems from two main sources: one is the software problem of making reasonable use of many arithmetic units; the other is the hardware problem of connecting the memory and the arithmetic units with a bandwidth that can match the speed of the processors.

These two difficulties are not independent of each other. A large part of the problem of using either workstations or MPPs comes from the fact that their memory bandwidth falls far short of the processing speed of the CPU. All modern microprocessors use the data cache to "hide" the access latency of the main memory. Though the latency is to some extent hidden, the bandwidth is not. Thus, to achieve an acceptable performance on modern microprocessors, we have to devise algorithms which exploit the data locality. This means that the data loaded into the cache must somehow be reused several times. In practice, we have to reuse the data at several different levels, from the registers, Level-1, -2 and, in some cases, -3 caches. To generate a program which can exploit the locality at all levels is a daunting task for both the application programmer and the compiler writer. With MPP, we also have one more level of the hierarchy, that is the memory of some other processor.

Actually, most MPP designers understand this problem well, and design machines with a relatively simple memory hierarchy. One of the extreme examples is the Fujituu VPP series, which uses the multiple-pipeline vector processor as the processing element. On this machine, the main memory has sufficient bandwidth to keep the arithmetic unit busy. On the other hand, CC-NUMA (Cache-Coherent Non-Uniform Memory Access) or COMA (Cache-Only Memory Architecture) machines rely heavily on deep memory

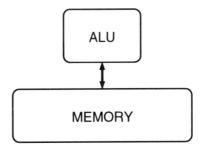

Figure 6   The von Neumann architecture.

hierarchy (see, for example, [LW95] for a detailed description and evaluation of a CC-NUMA architecture).

The basic difficulty comes from the von-Neumann architecture, which connects the arithmetic units and the memory directly (Figure 6). With this basic architecture, the bandwidth between memory and arithmetic units must increase as the speed and the number of arithmetic units increases. In practice, most computer designers come up with a memory hierarchy with the bandwidth diminishing at lower and lower levels. In theory, and in simulations on small problems, this architecture looks fine. Unfortunately, for many real applications it is very difficult to make effective use of a memory hierarchy.

The von-Neumann architecture is a general architecture which can be applied to any problem. Fifty years ago it made sense to develop such a general-purpose machine, since electronic computing devices were expensive to develop, no matter whether programmable or not. In fact, a fully-hardwired multiplier could not be implemented in a single machine until the late 1960s, and on an LSI chip until the late 1980s.

If we cannot build a hardwired multiplier, we must use some iterative algorithm to perform the multiplication. Thus, very roughly speaking, just to perform a multiplication we need programmable hardware. So it is quite natural to develop a programmable machine, since the efficiency we lose by adding the programmability is relatively small and the loss is more than compensated for by the wider range of applications, which amortizes the development cost.

However, as we have seen in the previous chapter, the evolution of the silicon semiconductor technology has resulted in a drastic increase in the number of transistors. Thus we can now pack tens of multipliers into a single chip, but we do not know how to construct a programmable machine which can use more than a few arithmetic units.

One way to solve this problem is to give up the programmability and design a machine dedicated to some limited class of problems. In the rest of this chapter, we briefly overview several successful attempts to develop special-purpose machines, and try to put them in a somewhat broader perspective.

3.2 Ising processors

3.2.1 Background

The Ising model is one of the most simplified models of physical systems consisting of many degrees of freedom. It consists of many spins organized on a lattice. Its Hamiltonian is written as

$$H = -J \sum_{<i,j>} S_i S_j - h \sum S_i, \tag{3.1}$$

where S_i is the value of the spin on the lattice point i. The first summation is over all nearest neighbor pairs; J is the coupling constant between the spins and h is the external field. Ising spin systems exhibit second-order phase transitions on a two or higher dimensional lattice. They are studied both theoretically and numerically as the simplest model with phase-transitions. Modified models, for example random bond or random field models, are used to study the spin glass, neural networks, proteins, and other many-body systems.

The Monte-Carlo technique is the most important method used to study these systems numerically. In the Monte-Carlo simulation, we select one spin, flip its value, and calculate the new energy. We then accept or reject the new configuration by comparing the energy change with some acceptance criteria. In the case of the Metropolis algorithm, we always accept the new configuration if ΔE is negative. When ΔE is positive, we accept the new configuration stochastically following the thermal fluctuation, so that the transition probability satisfies the detailed balance condition. Thus, the basic steps in an MC simulation of the Ising model are as follows:

- Select one spin in some way.
- Flip its spin and calculate ΔE.
- Accept the new configuration with the provability calculated from ΔE.

Calculation of ΔE can be implemented in very simple hardware, since each spin is represented by single bit. Random number generation requires larger hardware. In most cases, a shift-register based algorithm (sometimes called an LFSR, M-sequence or Tausworthe series) is used, since it requires no arithmetic operation. A more widely used algorithm to generate pseudo-random numbers on general-purpose computers is the multiplicative congruential scheme (see, e.g., [PTVF92]). However, for hardware implementation this scheme is inadequate, since the period of the sequence is determined by the size of the multiplier. To make the period acceptable, the size of the multiplier has to be very large. The shift register approach requires the amount of the hardware to be proportional to m, where the period is $2^m - 1$.

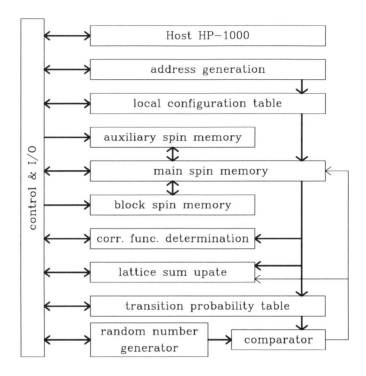

Figure 7   The Delft Ising Spin Processor.

3.2.2   DISP

DISP (Delft Ising Spin Processor [HSSC83], [HCB88]) was developed at Delft University of Technology in around 1982. The basic architecture of DISP is shown in Figure 7. It performed all the steps of the standard MC procedure in hardware. Since the standard algorithm allows only one site to be updated at a time, DISP updates only one site at a time. There are many algorithms that allow many sites to be updated independently. However, with the silicon technology available at the time DISP was developed, it would have been too expensive to develop a parallel machine.

DISP contains 4 Mbits of SRAM (64 64k-bit SRAM chips) to store spins. The basic Monte-Carlo step took 650ns on this machine. This cycle time is determined largely by the access time of the SRAM chip (250ns). On this machine, the address generator can handle both the sequential and random flips.

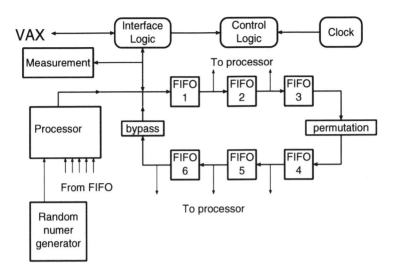

Figure 8   The Santa Barbara machine.

3.2.3   Santa Barbara machine

Pearson *et al.* [PRT83] developed a special-purpose computer for the Ising model based on a quite different architecture from that of DISP. In DISP, the memory access requires quite elaborate hardware, since it allows random flip. The Santa Barbara machine adopted a much simpler memory organization, by allowing only the sequential flip, in which the spins are flipped in a fixed order (sequential order). It only allows the skewed boundary condition because of the serial access to the memory. However, this architecture greatly simplified the design, since FIFO (First-In First-Out) memory can be used instead of RAM (random-access memory).

In addition, with the sequential flip algorithm it is relatively easy to implement a pipelined processor, since the order in which the data is accessed is fixed. Thus, the Santa Barbara machine operated on a clock cycle of 25 MHz, with the peak speed of 1 flip per cycle. Figure 8 shows a block diagram of the Santa Barbara machine.

3.2.4   Bell Lab machine

Condon and Ogielski [CO85] developed yet another special-purpose computer for Ising and spin grass systems. Its basic architecture is shown in Figure 9.

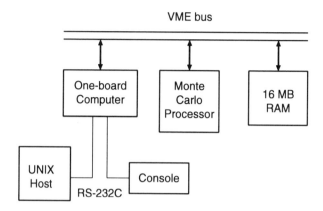

Figure 9   The overall structure of the Bell Lab Monte-Carlo processor.

A unique feature of this machine is that the spin memory is on the general-purpose VME bus, and it is shared by the special-purpose Monte-Carlo processor and a programmable one-board computer (MC68000 CPU). This one-board computer is connected to a console and a UNIX frontend through serial line.

This structure simplified the design task and reduced the development cost, since commercially available DRAM boards can be used without modification for the spin memory. The fact that the one-board computer can directly access the spin memory also simplifies the design.

The potential problem with this architecture is that the memory bandwidth is rather limited. However, for this machine it was not a serious problem. DRAM memory boards with a 240ns cycle time allowed a sustained performance of about 75% of the peak speed of 25 Mflips/sec.

Ogielski and Morgenstern [OM85] performed very large simulations of three-dimensional $\pm J$ Ising spin glass, and found evidence of the spin-glass transition. This was the first very important scientific result discovered with modern special-purpose computers.

3.2.5   *m-TIS and m-TIS-II*

These machines were developed by one of the authors (M. Taiji) in collaboration with M. Ito and several other people. The first hardware, m-TIS [TIS88], was a proof-of-the-concept machine built on an extremely small budget (essentially the pocket money of two graduate students).

Figure 10 shows the structure of m-TIS. The basic architecture of m-TIS is similar to that of the Bell Lab machine described in the previous section,

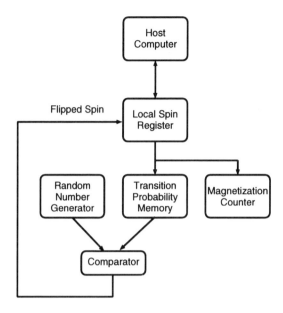

Figure 10 The overall structure of m-TIS.

but m-TIS is connected to the expansion bus of an Intel-based PC (80286 CPU) as a slave device. All the spins are stored in the host computer, and a part of them (16 spins) are transferred to the machine for simulations at one step. The Bell Lab machine is connected to the VME bus, and can work both as a master or slave device. When the CPU board accesses the Monte-Carlo processor, the Monte-Carlo processor operates as a slave device. During the actual simulation, the Monte Carlo processor operates as a master device and accesses the spin memory on the VME bus.

The advantage of having the master capability on the MC processor is that the bandwidth of the memory can be fully utilized. If the MC processor cannot access the spin memory directly, the CPU board must transfer the data between the MC processor and the spin memory, thus resulting in at least a factor of two loss in the memory bandwidth. This factor is quite important in the case of Ising machines, where the interaction is very short-ranged, compared with the case of the GRAPE systems described later.

On the other hand, to add the master capability on the MC processor significantly increases the design complexity of the MC processor, and it also limits the flexibility of the machine, since the MC processor can access the memory only in the way implemented in its hardware.

For m-TIS, the designers made the choice to design the special-purpose computer as a slave device. This simplified the design quite significantly: it took only four months, including both the design and fabrication. The architecture also made it possible to apply the system to a wide range of problems without changing the hardware. In the case of the Bell Lab machine, it was necessary to reprogram the PLD (Programmable Logic Device) chips to change, for example, the size of the lattice to be simulated. With m-TIS, such a change was not necessary since the programmable host computer took care of the address generation. It is also easy to change lattice structures.

The m-TIS II hardware pioneered the use of the FPGA as a custom computing device. All of the core logics of m-TIS II were implemented in three Xilinx 3090 devices. As a result, the user can change almost anything, such as the algorithm to generate the random numbers, the topology of the lattice, and so on.

3.3   Caltech Hypercube and other QCD machines

3.3.1   Background

The Lattice Quantum Choromo-Dynamics (Lattice QCD) calculation has been regarded as one of most compute-intensive problems in pure sciences (see [FWM94] and the references therein).

As is the case with many other problems in computational physics, the speed requirement for the computer in QCD calculations always by far exceeds what was available in general-purpose computers. For QCD simulations, it was not exceptional to dedicate a large supercomputer to a single simulation for several months or even years. Thus, things like ease of use are not the first priority.

Readers who are interested in details of the Lattice QCD calculations are referred to [FWM94], which gives an overview of QCD calculation as well as a number of references. Here, it is enough to say the QCD calculation is a Monte-Carlo calculation on a 4-dimensional grid, since existing QCD machines use only this fact as the design guideline.

3.3.2   Caltech machines

The Caltech parallel computer group [FWM94], [Fox88] is one of the most successful research efforts in computational science, resulting in research that covers almost the entire range of parallel computing, not just computational physics or scientific computing. The project started as an effort to develop a cheap parallel computer based on the 8086 microprocessor to run QCD calculations in 1981.

The architecture chosen was a parallel processor made of programmable computers (Processing Elements, PEs). Each PE has a central processor, a local memory and communication interfaces. In other words, each PE is a complete programmable computer. In early machines, PEs are connected in

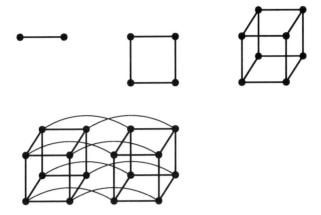

Figure 11   Hypercubes for up to $k = 4$.

Table 3   The Caltech machines.

Machine	Year	Number of processors	CPU	Peak speed (PE/total)
Mark-I	1983	64	8086/8087	50k/3M (est)
Mark-II	1984	128	8086/8087	50k/6M (est)
Mark-III	1986	128	68020/68882	0.2M/60M (est)
Mark-IIIfp	1989	128	68020/Weitek	16M/1.9G
Intel Delta	1991	512	i860	60M/32G

a k-dimensional hypercube, where 2^k is the number of PEs. This architecture has the advantage that the distance between any two PEs in the system does not exceed $k = \log_2 p$, where p is the number of PEs. On the other hand, the hardware cost for the communication network grows as $p \log_2 p$. Table 3 gives the specifications of the Caltech machines and the Intel Touchstone Delta. The Delta adopted the 2-D mesh network with wormhole routing.

Though QCD calculation was the initial target, the Caltech machines were quickly accepted as high-performance, multi-purpose parallel computer resulting in several commercial versions, such as the Intel iPSC and nCUBE.

As far as raw performance is concerned, the achievement of the Caltech machines is less than impressive. They never outperformed top-of-the-line supercomputers. All Caltech machines used commercially available chips for both the CPU and FPUs.

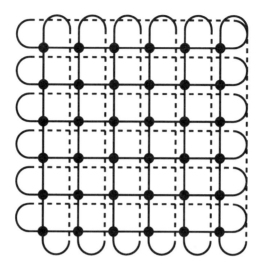

Figure 12   A 6×6 2-dimensional torus network.

3.3.3   PAX project

The PAX project [Hos92] started as a project to make an inexpensive computer that could run simulations of systems described by a set of partial differential equations. Thus, their original goal was broader than that of the Caltech group. They chose a distributed-memory computer as the basic architecture, and a 2-D torus connection as the communication network. Thus, the network topology was simpler than that of the early Caltech machines. The reason they chose the 2-D torus is that, for a practical number of processors, the 2-D torus is close to optimal, if the trade-off between factors such as the difficulty in packaging are taken into account. They demonstrated that communication between the processors was rarely the most time-consuming part for most of the problems they solved on their several generations of PAX machines. The Caltech group seems to have reached the same conclusion with their Delta machine.

In PAX machines, the communication link between the neighboring processors is implemented as a small memory physically shared by two processors.

In 1977, the first machine, PACS-9 (at this time, the acronym PACS stood for Processor Array for Continuum Simulation) was built. Table 4 gives a summary of the project.

All of the machines, except for CP-PACS, were designed under the leadership of Professor Hoshino, resulting in very simple and efficient hardware designs. The last three machines were all marketed as general-purpose parallel computers, though not too many of them were actually sold.

Table 4   The PAX machines.

Machine	Year	Number of processors	CPU	Peak speed (PE/total)
PACS-9	1977	9	6800	1k/10k
PACS-32	1980	32	6800/Am9511A	16k/0.5M
PAX-128	1983	128	68B00/9511A-4	30K/4M
PAX-64J	1984	64	LSI-11	0.1M/6M
QCDPAX	1989	432	68020/L64133	28M/12G
CP-PACS	1996	2048	extended HP-PA	300M/614G

CP-PACS diverted from its predecessors quite drastically, adapting the 3-dimensional hyper crossbar as the communication network and paying far more attention to the software than was done on previous machines. As a result, CP-PACS has become a relatively easy-to-use, but rather expensive, machine (2.2 B JYE for the peak speed of 600 Gflops as delivered by late 1996), compared to its predecessors.

Machines up to QCDPAX adapted commercial chips as both the CPU and floating-point operation units (in other words, they used COTS — Commodity, Off-The-Shelf technology — well before that phrase was coined). For QCD-PAX, a custom LSI was used to interface the CPU and FPU. For CP-PACS, a custom processor based on the HP-PA architecture is used to effectively achieve the vector processor capability without adding the complete set of vector instructions (pseudo vector processor). CP-PACS is also quite different from its predecessors in the manufacturing technology used. The processor chip is fabricated using 0.3μm technology. The number of transistors per chip is 4.5 million, and the package has more than 1000 pins (more than 500 used for actual signals).

3.3.4   Columbia machines

The Columbia group completed their first machine in 1985 [Ter88]. It consisted of 16 PEs, each of which has one Intel 80286/80287 microprocessor and a TRW 22-bit floating-point processor. Each PE had a speed of 16 Mflops. In 1987 they completed the second machine (64 PEs, total speed of 1 Gflops). The Weitek chips are used for this second-generation machine. In 1989 they completed the third machine (256 PEs, 16 Gflops), also using Weitek chips. They are currently developing their fourth machine, which will consist of 16 384 PEs connected in a 4-D grid (QCDSP). It will have the peak speed of 0.8 Tflops. The anticipated total cost of this machine is $ 3M. The QCDSP use the TI

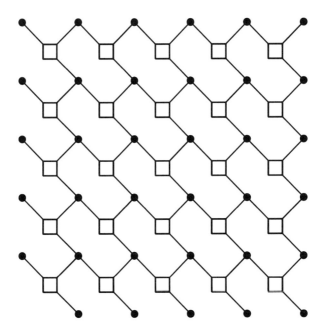

Figure 13 The structure of the 2D grid used in early Columbia machines.
Filled circles denote the processors and squares denote memory units.

32C030 DSP chip as the PE (hence the name QCDSP). The TI chip operates
at a clock speed of 25MHz and offers speed of 50 Mflops.

Early Columbia machines have an interesting architecture (shown in Figure
13). The processors form a two-dimensional torus network in the same way as
in PAX machines, but the neighboring processors share the local memory. In
early PAX machines, each node has the main memory unit and a small memory shared with the neighbor. With QCDSP, however, the Columbia group
has also changed the communication structure completely, and adopted a four-
dimensional grid. Each connection is a bit-serial interface with a data rate of
50 Mbps. Thus, the number of wires is not excessive. Since the number of PEs
for QCDSP is very large, a 4-D network is preferred over lower-dimensional
topologies.

3.3.5 APE machines

The first of the APE machines was also completed around 1984, with a peak
speed of 256 Mflops. It has a rather complex architecture of a circular ring with
the memory-processor connection network, which allows the circular shift.

Their second generation machine, the APE-100, became operational around 1993. APE-100 is quite different from other QCD machines in the sense that it is a true SIMD machine without even a local addressing capability. As of early 1996, the largest APE-100 machine actually in operation seems to be that in Rome, with 1024 PEs and a peak speed of 50 Gflops. APE-100 is also marketed as Quadrics. APE-100 used custom-design ASIC chips for the floating-point operation units. A single node of APE-100 operates at a 25MHz clock speed and can perform one addition and one multiplication in single precision. There is no hardware support for double precision. The node ASIC is fabricated with a 1.2μm process, and had the die size of less than 1 cm^2.

The third generation machine, the APEmille, is reportedly now under development for a total grant of \$ 13M. APEmille retains the basic SIMD concept and three-dimensional interconnects. The processor will operate at 50 $\sim$ 100MHz and perform two additions and two multiplications. Thus the speed of one node will be 8 $\sim$ 16 times the speed of the present APE-100, which is sufficient to achieve the planned peak speed of around 1 Tflops.

3.3.6   The GF-11

The GF-11 project was started in the early 1980s. The machine became operational by late 1988 $\sim$ 1989. It is an SIMD multiprocessor with 566 PEs, and has a unique architecture in which the memories and PEs are connected through a 2-stage Omega network made of 24-by-24 full crossbar switches. Thus all processors can access any memory location with uniform latency and bandwidth. It uses the Weitek 1264/1265 chipset as the floating-point operation unit.

3.3.7   QCD machines and commercial MPPs

All QCD machines except the GF-11 are essentially distributed-memory parallel computers with a relatively simple communication network. Thus, they are quite similar to commercial MPPs. In fact, many of the commercial MPPs are originated from QCD machines. To some extent, commercial MPPs are the result of the success of the QCD machines for applications other than QCD calculations.

However, this success did cause a quite fundamental problem. Commercial MPPs have become just as good as custom-made machines, since the architecture is the same. Commercial machines have the very important advantage that you do not have to build them yourself. Thus, you can use the rare human resources for other things, like developing software, running simulations and writing papers, or in short, to do science. In addition, commercial machines have a cost-performance comparable to that of custom-made machines, since the former can amortize the development cost over many machines.

In the early 1980s, when the fastest commercial computers were vector machines, to develop a distributed-memory parallel computer for QCD calculation was an effective way to achieve a cost-performance better than that of commercial machines. However, since present commercial high-performance computers and QCD machines are essentially identical, there is no real merit in building it yourself. Fox [Fox88] made a similar argument, and stated that it is unlikely that the Caltech group would continue the hardware development.

If we look into the detailed design of the hardwares, however, we still see many differences. For example, both the Columbia machine and APE use 32-bit arithmetic, instead of the 64-bit arithmetic used in almost all general-purpose computers. Though most modern computers can handle both 32- and 64-bit data, they handle it at the same speed. Thus, half of the memory bandwidth and more than three-quarters of the multiplier circuit are not used in 32-bit arithmetic. One can thus gain at least a factor of two by designing the machine for 32-bit operation.

In addition, the balance between the memory size and the processor speed tends to be quite different. Many general-purpose computers apparently have the memory size roughly proportional to the speed of the processor, seemingly because many applications require an almost linear relationship between the memory size and the calculation speed. One example is three-dimensional computational fluid dynamics. If we use an explicit method, the memory requirement is $O(L^3)$, where L is the number of grid points in one dimension. The calculation cost is $O(L^4)$, since the number of timesteps is proportional to $1/L$ from the Courant condition. Thus, we have the relationship

$$\text{(memory size)} \propto \text{(speed)}^{3/4}. \tag{3.2}$$

This is not very far from a linear relationship. Thus, most high-performance computers have been designed with "1 Gbytes of memory for 1 Gflops".

The "scalable" machines have made this linear relationship between the memory size and speed difficult to avoid. These machines are designed to be usable in relatively small configurations like 4 or 8 PEs, since most customers buy small configurations. Thus, PEs are designed to have a relatively large memory, and the system software tends to eat up a sizable fraction of the memory available on a typical configuration. Larger configurations are thus forced to have a very large memory.

The QCD calculation does not require that much memory. Thus, QCD machines currently under development typically have the memory significantly smaller than that of commercial computers of a comparable speed.

Therefore, even if the architecture is the same as that of general-purpose MPPs, a machine optimized for QCD calculation can have a significant gain. The Columbia 0.8 teraflops machine is the most aggressive QCD machine currently under development. It will be completed soon, for a total cost of \$ 3M. This is at least a factor of 10 better than what is available in the form of a commercial computer.

A factor of 10 advantage in cost-performance is more than enough to justify the effort. As we have seen in Chapter 1, the speed of general-purpose computers increases by a factor of 10 in five years. Thus, if performance better by a factor of 10 is achieved, it gives an advantage of five years.

3.4 Molecular dynamics machines

3.4.1 Background

Molecular dynamics is a technique used to study the properties and behavior of liquids or solids at the molecular level, by integrating the trajectory of each atom in the system. A macroscopic object typically consists of $> 10^{24}$ atoms, which is far too large to simulate on any conceivable computer. Fortunately, many important results can be obtained with a much smaller number of atoms.

In classical molecular dynamics simulations, the interaction between atoms is modeled by a simple potential function obtained either theoretically or experimentally. There are three different kinds of interaction. The first is interaction through chemical bondings, the second is the van der Waals force between neutral, non-bonded atoms; and the third is the Coulomb interaction.

In terms of the calculation cost, chemical bonds are the least expensive, since the total number of bonds is typically a few times the number of atoms, N. The van der Waals force is quite a bit more expensive. The Lennard–Jones potential is the most commonly used model for the van der Waals force: it has the form

$$\phi = ar^{-12} - br^{-6}. \tag{3.3}$$

Thus, at a large distance it falls off as r^{-6}. If we want to achieve, say, 0.1% accuracy for the force, we have to take into account the forces from atoms about three times the distance to the nearest atom. The number of atoms in this range is around 100.

The Coulomb force poses a difficult problem. Its potential falls off only as $1/r$. Thus, strictly speaking, to calculate the Coulomb force on an atom, we have to take into account the interactions with all other atoms in the system. This means that the calculation cost is $O(N^2)$ (see Figure 14). In practice, an artificial cutoff is used in most cases to reduce the calculation cost. Theoretically, the cutoff might be justified if the system is electrically neutral in the scale smaller than the system size. In some cases, the Ewald method [Ewa21] is used to calculate the force in a periodic boundary condition without cutoff.

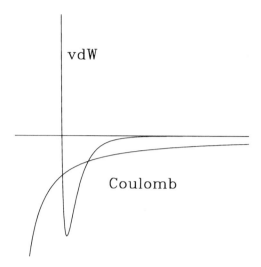

Figure 14   The Coulomb and Lennard–Jones (van der Waals) potentials as a
function of the distance. The van der Waals potential approaches zero quickly
for a large distance.

In the case of simulating a system which consists of neutral atoms, calculation of the van der Waals force is the most expensive part. Two algorithms
are used to calculate the van der Waals force, one of which is the neighbor
list method. In this method, each atom maintains a list of its neighbor atoms.
To obtain the force on a particle, only the forces from atoms in this list are
calculated. The neighbor list is updated at some fixed time interval, either
by scanning over the whole system or by using the other method, described
below. This method is fairly simple to implement, and the calculation cost
is close to the theoretical minimum. A major disadvantage of this scheme is
that it requires quite a large amount of memory.

The other method, sometimes called the linked-list method, is based on grid
subdivision of the space. In the simplest case, if the effective distance of the
force is r_c, we divide the total simulation volume to the cubic cells with a
size not smaller than r_c. The force on an atom is calculated by summing the
forces from atoms in the cell in which it resides and atoms in 26 neighboring
cells. This method does not require much memory, but the calculation cost
is larger by a factor of six in the case of the above simple implementation.
This is because of the overhead of calculating the forces from the atoms that
are actually outside the sphere of radius r_c. This overhead can be reduced by
using a smaller cell and considering the cells next to the direct neighbors.

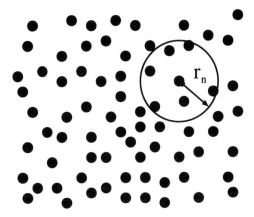

Figure 15   The neighbor list algorithm. Only the forces from atoms within the radius r_n are calculated.

3.4.2 DMDP

DMDP (The Delft Molecular Dynamics Processor) [BB88] is a machine specialized for molecular dynamics simulations. Figure 17 shows the basic structure of the machine. It implements all simulation procedures, including the analysis of the result, in hardwired logic. At the time of completion (around 1980), its performance was truly impressive. It offered a peak speed comparable to that of Cray-1, and achieved a similar speed as Cray-1 for real applications, for a development cost of less than 1% of the price of a typical Cray-1 configuration.

The key technology in achieving this extremely high price-performance is the pipeline to calculate the force between particles. In molecular dynamics simulation, we integrate the equations of motions of many atoms which interact with the rest of the system. The main target of DMDP is the system of atoms which interact through short-range forces (the Lennard–Jones force). Thus a typical atom interacts with a few hundreds atoms. To select the particles which interact with each other, the linked-list scheme was implemented in hardware. The particles that interact are sent to the hardwired pipeline to calculate the force between them, and the velocity vectors of these particles are directly updated using the calculated pairwise force. It has no separate memory for the acceleration. This organization reflects the standard algorithm in which the leapfrog (Verlet) integrator is implemented.

The application range of DMDP is rather limited since it cannot change either the integration scheme or the type of interaction. The hardware to calculate the interparticle force is programmable, since it calculates the force

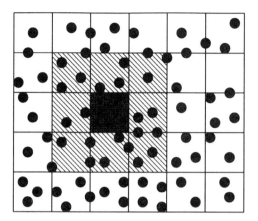

Figure 16   The linked-list algorithm shown in two dimensions. Forces from atoms in nine cells (shaded) are calculated to obtain the forces in the cell shown in gray.

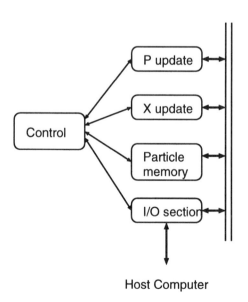

Host Computer

Figure 17   The overall structure of the DMDP.

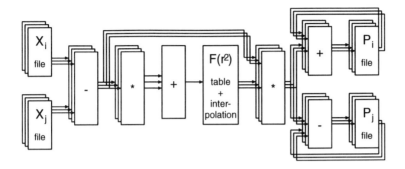

Figure 18   The momentum update hardware of DMDP.

by lookup table and linear interpolation. However, it can calculate only one kind of interaction at a time. Thus, it is fairly difficult to use DMDP for the simulation of anything more complicated than single-atom molecules such as Neon or Helium.

The basic design of DMDP was completed in 1978, and the hardware was completed by 1982. The basic building blocks of the machine are 16 Kbit DRAM chips and 8-bit multipliers. The total cost of the machine was 150 000 Dfl.

Figure 17 shows the basic structure of the machine, which consists of the pipeline to update momentum, the pipeline to update position, the memory and I/O sections. The memory consists of three memory modules for position data (24 bits each), another three for momentum data (32 bits each), one unit for the linked list (16 bits) and one unit for particle type descriptor (for extension). Each module has its own dedicated bus, thus the data bus has a width of 188 bits.

The part to update the position is just a Multiply-And-Accumulate (MAC) unit to perform the operation

$$x \leftarrow x + cP, \tag{3.4}$$

where the constant c is defined as $c = \Delta t/m$. Here Δt is the timestep and m is the mass of the atom.

Figure 18 shows the momentum update part, which implements the calculation

$$\Delta \mathbf{P}_i = \sum_j F(r_{ij})(\mathbf{r}_i - \mathbf{r}_j). \tag{3.5}$$

This is by far the most expensive part of the calculation, and the hardware for the momentum update is the largest part of DMDP.

The basic idea of DMDP is to accelerate this operation by using a large amount of hardware. In order to achieve this goal, a completely hardwired pipeline which can process one interaction per clock period is constructed. This hardware calculates all the interactions between the atoms in two cells, and updates the momentums of these atoms. For one timestep, this hardware is used for all pairs of neighboring cells.

The position data are stored as 24-bit fixed point numbers. After the subtraction $x_i - x_j$, the result Δx is shifted so that the next multiplication to calculate Δx^2 does not cause any overflow/underflow for the region where the force needs to be calculated. Note that, for the target application of DMDP, the force can be truncated at a finite distance several times larger than the radius of atoms. In addition, it is impossible for two atoms to become closer than, say, half of the radius, since the potential diverges very quickly. Therefore, the force needs to be defined only for the range of r of about a factor of 10.

The squared relative distance Δr^2 is obtained by adding the three components. The upper 10 bits of Δr^2 is used as the entry for the force lookup table. This table produces the zeroth order value in 32-bit and first order value in 16-bit format. The first order value is then multiplied with the lower 16 bits of Δr^2, and the result is added to the zeroth order value to perform linear interpolation.

The result, $f(r)$, is then multiplied with Δx, etc. to obtain the three components of the force. This force is directly added to $\mathbf{P}_i$ and subtracted from $\mathbf{P}_j$. The system of units is chosen so that we do not need to multiply the force with the timestep.

The force calculation part operates at the cycle time of 250 ns, resulting in a peak speed of around 150 Mflops.

The DMDP system consists of around 20 (estimated from the description in [BB88]) wire-wrapped boards, each of which has a size of 37×28 cm (9U Eurocard). These boards have three 96-pin connectors, and are all connected to the backplane bus.

The speed of 150 Mflops achieved in the early 1980s is quite impressive. It roughly corresponds to 100 Gflops of the mid-1990s. Thus, as far as the performance achieved is concerned, DMDP should be regarded as the principal demonstration of the superiority of a special-purpose design over general-purpose computers.

Bakker, however, did not continue the development of machines of a similar architecture. The next machine he built for molecular dynamics, the ATOMS [BGGT90], differed quite drastically from DMDP in its basic design, becoming more like a usual programmable multicomputer so that it could handle more complex kind of interactions such as multi-body effects.

3.4.3   FASTRUN

FASTRUN [FDL91] is a special-purpose computer designed to simulate protein molecules. It consists of two parts: one is a custom-made machine to calculate the Coulomb and van der Waals forces by a hardwired pipeline; the other is a general-purpose array processor designed to do all the other tasks. It has very elaborate hardware/software to implement the neighbor list algorithm and to calculate the truncated Coulomb force with reasonable efficiency.

The pipeline to calculate the force between particles is similar to that of DMDP, but quite a bit more complex. For example, it calculates both the Coulomb and van der Waals forces and potentials simultaneously. Thus, the pipeline contains four independent function evaluators, each of which performs the table lookup and quadratic interpolation. In addition, there is a fifth interpolation unit which calculates the distance r from the square of the distance r^2. The numerical accuracy is improved over DMDP, since FASTRUN uses single-chip floating-point chips from Weitek as the basic building blocks.

The FASTRUN pipeline operated at a cycle time of 220 ns, only slightly faster than that of DMDP.

3.5   The Digital Orrery

The Digital Orrery is a machine dedicated to the long-term integration of the orbits of planets. Its architecture is that of an SIMD computer with a one-dimensional ring network. Each processor (planet computer) calculates the gravitational force on one particle and the integration of its orbit.

The planet computer was designed around an experimental 64-bit floating-point chipset developed by Hewlett Packard. Each chip can perform one floating-point operation in 1.25 μs. A planet computer consists of this chipset and two memory units, and the datapath to connect these units. The controller sends instructions to planet computers at each machine cycle.

This architecture is well suited for the long-term synchronous integration of small-N systems. For example, the 14th order Störmer method costs about 200 floating-point operations. The calculation cost of other high-order schemes is not much different. Thus, the calculation cost of the gravitational force and that of the time integration are comparable. Moreover, we need to use a word length longer than 64 bits for the position data to ensure accuracy. Thus, the calculation cost of the time integration was perhaps larger than the cost of the force calculation.

Time integration using the Störmer method requires a relatively high memory bandwidth. The Störmer method is the usual linear multistep method for the second order equation, which is expressed as

$$x_{i+1} - 2x_i + x_{i-1} = h^2 \sum_{k=0}^{m} C_k f_{i-k}. \tag{3.6}$$

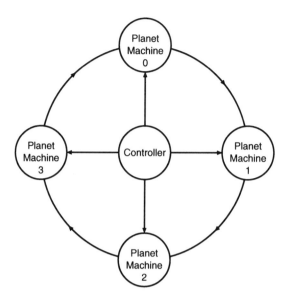

Figure 19   The Digital Orrery. The actual number of planet machines is 10.
Only four are shown here.

Thus, the ratio between the memory access and the floating-point operation is about 0.5. In other words, the hardware must be designed so that at least one memory unit exists for each of the floating-point units. The SIMD architecture of the Orrery is close to ideal for the parallel integration of the orbits of planets.

At present, the SIMD architecture is not very popular. One reason is that it is practically impossible to achieve a reasonable speed with a classical SIMD architecture. In an SIMD architecture, the central sequencer broadcasts instructions to all processing elements. In addition, usually the whole system operates on a single global clock. If the system is physically large, it becomes very difficult to increase the frequency of the global clock. However, at the time of the design of Orrery, this was not a severe limitation, since the system clock frequency was limited by the speed of the floating-point chipset and not by the speed of the instruction bus.

The Orrery was quite successful as a dedicated machine for the long-term integration of the outer planets. Sussman and Wisdom [SW88b] were the first to find that the orbit of the planet Pluto is not stable. The orbit is unstable if the distance between two solutions from slightly different initial conditions increases exponentially. They integrated the orbits of Jupiter through Pluto, and found that the orbit of Pluto is unstable and could be chaotic. It took several years after their initial work for confirmation of their result to appear in the literature.

As a follow-on to Orrery, Sussman and his collaborators developed the "Supercomputer Toolkit", which is an MIMD message-passing style multicomputer with a reconfigurable (by actually changing the physical connection) communication network.

3.6 Summary

In this chapter, we have overviewed special-purpose computers for scientific simulations. Several machines for Monte-Carlo simulations of Ising and related models in statistical physics were built in the 1970s and 1980s. These machines achieved superior performance compared to contemporary general-purpose computers because the size of the circuit (in other words, the number of part counts) was orders of magnitudes smaller. In the late 1980s and 1990s, however, it has become increasingly more difficult to achieve high performance with specialized computers for Ising models, because the small circuit size has ceased to be an advantage.

In the 1970s, a small circuit size was a significant advantage because it directly determined the cost of a machine. In the early 1970s, the building block of computers was Small-Scale Integrated (SSI) circuits with around 100 transistors. A general-purpose computer comprised thousands or millions of these IC chips. An Ising processor with a similar performance to that of a large general-purpose computer could be constructed with hundreds of IC chips, resulting in a huge difference in the building cost.

In the late 1980s, however, a single LSI could accommodate millions of transistors, and a personal computer with a CPU made by such technology costs less than \$ 10 000. One could still build an Ising machine for a few hundred dollars, but if the development time of the hardware and software is taken into account, using a general-purpose computer would be more productive.

It was also possible to design a custom processor chip for the Ising model. However, its performance would be limited by the off-chip communication bandwidth, and the relative advantage over the general-purpose computer would not be as large as that in the 1970s.

The machines for Lattice QCD simulations followed a path quite different from that of the Ising processors. Instead of constructing a processor specialized for a specific part of the calculation, the designers of QCD machines took single-chip microprocessors and floating-point arithmetic units, assembled a simple computer by adding memory, and connected many of them in some form of network. This approach was novel in the early 1980s. However, in the late 1980s and 1990s, general-purpose massively-parallel computers followed the design of QCD machines, since the design turned out to be applicable to a much wider class of problems than originally anticipated.

Ironically, the relative advantage of QCD machines over commercial general-purpose computers, as the result of their very success, diminished over the last 10 years, and many of the projects to build QCD machines also target the commercial market outside QCD.

Several machines were built for the classical many-body problem in molecular dynamics or celestial mechanics. They did achieved impressive performance, and some were quite successful in obtaining important scientific results. However, there were no follow-ons of a similar architecture for any of these machines. This is quite different from the QCD machines, where several groups have been developing machines for nearly two decades.

It is difficult to understand why there were no follow-ons, but DMDP and Digital Orrery seem to be saying something. The designers of these machines did build new machines, but with these architectures similar to general-purpose multiprocessor or QCD machines. This suggest that the designers thought the specialized architectures of DMDP or Orrery were too problem-specific and not sufficiently useful to justify the development of follow-on hardware. The decision of these designers is quite understandable, as far as the overall usability of the machines is concerned.

In the next chapter, we describe our GRAPE project, yet another project to build a special-purpose computer for the many-body problem. We discuss how we solved the above difficulty of limitation in the application range.

4

The GRAPE Systems

4.1 Prehistory

The GRAPE systems are specialized to the time integration of gravitational N-body systems. The gravitational N-body system is expressed simply as the collection of many particles (stars) interacting dominantly through gravity. One particle in the gravitational N-body system feels the gravitational forces from all other $N - 1$ particles in the system.

The important characteristic of the gravitational N-body system which makes it quite unique is that gravitational force can reach an infinite distance. The strength of the force is proportional to the inverse of the square of the distance r. If we consider the force on a particle in, say, a globular cluster with 10^5 stars, a large part of the force comes from particles far away from it. Thus, we need to calculate the force from the entire system. This is quite different from the van der Waals force between atoms, which goes off as r^{-7}. The Coulomb force between atoms has the same r^{-2} dependence as the gravitational force. However, effectively it goes off more rapidly because the whole system is electrically neutral. Thus, the monopole vanishes and the force falls off at least as r^{-3}.

The most straightforward way to handle the gravitational N-body system is to calculate directly the force on each particle as the sum of the forces from other $N - 1$ particles. GRAPE is designed essentially for this purpose.

Figure 20 shows the most simplified view of a GRAPE system. A GRAPE system consists of two components, the host processor and the GRAPE processor. The host processor performs all calculations other than the force calculation, and GRAPE processor performs the force calculation. The host processor sends the positions and mass of the particles to the GRAPE processor, and the GRAPE processor calculates the forces on particles.

This idea of building a special-purpose computer for N-body simulation is not exactly an original one. In summer 1988, Professor Chikada, then of Nobeyama Radio Observatory, gave a talk on the general possibility of building a special-purpose computing device for astrophysical simulations. Just

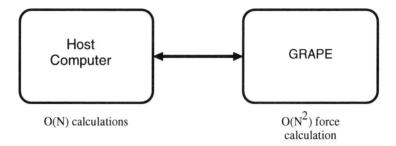

Figure 20   Basic structure of a GRAPE systems.

as one example, he described the pipelined hardware to calculate the gravitational force between particles. He described a force calculation pipeline similar to that of DMDP, with a speed of 300 Mflops (10 MHz clock). For historical interest, we reproduce his original description in Figure 21. He estimated the construction cost would be around four million Yen. The cost performance would be more than 100 times better than that of general purpose computers available at that time. Even at the time of writing this book (1997), a sustained speed of 300 Mflops would cost a few million Yen.

Chikada was famous for his FX computer [CIH⁺87], specialized for image synthesis from a radio interferometer. FX is essentially a huge special-purpose computer to perform low-accuracy FFT operations at a speed of 100 GOPS, which cost \$ 1M in the early 1980s, when a vector processor with a peak speed of 0.4 Gflops would have cost more than \$ 10M. Chikada thought a similar approach would be also useful for simulations.

He distributed the summary of his talk to many theoretical astrophysists in Japan. Sugimoto, then the thesis advisor of one of the authors (J.M.), was among those who received Chikada's summary, and the only person to respond it. At that time J.M. and Sugimoto had been working on N-body simulation of the post-collapse evolution of globular clusters for several years (see Section 6.2). This kind of simulation demands huge computing power, since the calculation cost is proportional to $N^{3.3}$ [MH88],[HMM88]. The calculation cost per one timestep is $O(N^2)$. Another power of N comes from the fact that the evolution timescale of the system is proportional to N. The last $N^{0.3}$ comes from the size of the timestep. The timestep must be small enough to accurately follow the change of forces from nearby particles. Therefore, the timestep is proportional to the average interparticle distance, which is of the order of $N^{-1/3}$. By the end of the 1980s, the largest number of particles ever tried was 3000. This calculation took several hundred CPU hours on a Cray X-MP/18 [Mak89].

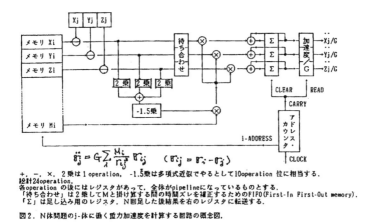

$$\ddot{\mathbf{r}}_j = G \sum_i \frac{M_i}{r_{ij}^3} \mathbf{r}_{ij} \quad (\mathbf{r}_{ij} = \mathbf{r}_i - \mathbf{r}_j)$$

+, -, ×, 2乗は1 operation, -1.5乗は多項式近似でやるとして10operation 位に相当する。
総計24operation.
各operation の後にはレジスタがあって、全体がpipelineになっているものとする。
「待ち合わせ」は2乗してMと掛け算する間の時間ズレを補正するためのFIFO(First-In First-Out memory).
「Σ」は足し込み用のレジスタ. N回足した後結果を右のレジスタに転送する.

図2. N体問題のj-体に働く重力加速度を計算する回路の概念図.

Figure 21   Chikada's illustration of the pipeline processor for gravitational
interaction between particles.

More approximate simulation of globular clusters such as the one-
dimensional conducting gas calculations [SB83],[BS84],[Goo87] and the direct
integration of the orbit-averaged one-dimensional isotropized Fokker–Planck
equation [CHW89] have shown that the behavior of the system would be quite
different for $N < 8000$ and $N > 8000$. In short, systems with small N are
stable, while large-N systems are unstable and exhibit oscillation of the central density. The amplitude of the oscillation is larger for larger N, and for
sufficiently large N, the oscillation becomes chaotic. We give more detailed
discussion in Chapter 6.

Because of the many simplifying assumptions used in these calculations,
whether such an oscillation could actually occur in real globular clusters was
not clear. To obtain a definite answer, it is necessary to perform N-body
simulations with a sufficiently large number of particles. So we investigated
the possibilities of fast calculation using various kinds of parallel computers
and vector processors.

What we found was that distributed-memory parallel processors such as
CM-2 [Hil85] were not well suited for the kind of time-integration algorithm
we were using, though they offered a much better price performance than
that of traditional vector processors. On the other hand, the vector processors
were not fast enough, though they were well suited for the algorithm we were
using. In practice, vector machines and parallel processors showed similar
performance [MH89].

The main reason that parallel processors are not well suited for our problem is that the number of particles we can handle is small, and increases very slowly as we increase the speed of the computer. Even with a speed of 1 Teraflops, the number of particles we can handle in a realistic time is less than 10^5. On the other hand, any conceivable teraflops system built using present or near-future technology would consist of far more than 1000 processors, since otherwise the clock frequency must go beyond 1 GHz. In other words, one processor holds only a few tens of particles. With that small number of particles per processor, the interprocessor communication will dominate the total calculation time, unless the communication network offers extremely high bandwidth and low latency.

Another problem we found was that both the vector processors and parallel processors had too much memory, and the memory cost seemingly determined the total cost of the machine. As discussed in Chapter 3, for many important problems, the calculation cost is $O(N^\alpha)$ with α close to one. Thus, as the speed of the computers increases, the size of the problem, in other words, the amount of memory to store the variables, grows in proportion to the speed. For our problems, however, the calculation cost is as high as $O(N^{3.3})$. Therefore, even for the teraflops speed, the amount of memory we need would be less than 100 MB. On the other hand, a 1 Gflops computer typically has 1 GB of memory. For a speed of 1 Gflops, the actual amount of memory we need is less than 10 MB. Therefore, high-performance computers have several orders of magnitude more memory than we need.

For us, therefore, the proposal by Chikada was particularly attractive. Before we saw Chikada's proposal, we did think of developing something along these lines. We had heard about the Digital Orrery (Section 3.5), so we knew that someone with the necessary knowledge could build a special-purpose computer for the N-body problem. However, we had not actually started to design/build any hardware, essentially because we had no experience or knowledge of designing hardware. We had no idea how difficult it would be. Chikada's proposal stated it is not very difficult. For us, Chikada's proposal meant that we would have someone to consult on the design and development of the hardware, which was the very thing we needed to start the project.

By late 1988, Sugimoto decided to organize an effort to develop the hardware for gravitational force calculation. Tomoyoshi Ito, who just passed the entrance examination of the graduate school, joined us. Until April 1989, when the new academic year started, he was still busy with other things, so we used this time for preparations such as studying the basics of the digital logic and acquiring minimal equipment like an oscilloscope and a circuit tester. By June 1989, Toshikazu Ebisuzaki joined our group as a research associate. These four people were the initial members of the GRAPE project. The name GRAPE itself was proposed by Sugimoto in summer 1989. Until then, our project has no name.

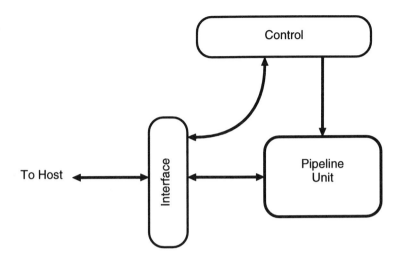

Figure 22   The GRAPE-1 system.

4.2   GRAPE-1

GRAPE-1 [IMES90] was the first hardware we ever designed/built. It was originally intended not as a system useful for real work, but as a design/development exercise. The calculation performed in GRAPE-1 hardware is the evaluation of the following summation:

$$\mathbf{a}_i = \sum_{j=1}^{N} \frac{\mathbf{x}_j - \mathbf{x}_i}{(r_{ij}^2 + \epsilon^2)^{3/2}}, \tag{4.1}$$

where $\mathbf{a}_i$ is the gravitational acceleration at the position of particle i, $\mathbf{x}_i$ is the position of particle i, r_{ij} is the distance between particles i and j, and ϵ is the artificial potential softening used to suppress the divergence of the force at $r_{ij} \to 0$.

Figure 22 shows a block diagram of GRAPE-1. It consists of the interface unit, the pipeline unit and the control unit. The interface unit handles the communication with the host processor. The pipeline unit performs the actual force calculation. The control unit generates control signals for the other units.

Figure 23 shows the basic structure of the pipeline unit. The overall structure of the pipeline itself is similar to that of the DMDP (Figure 18). The only difference is that this pipeline accumulates the force on one particle, while the momentum update pipeline of DMDP updates the momentum on two particles, using the symmetry of the force. For GRAPE-1, we decided not to use the symmetry to simplify the hardware.

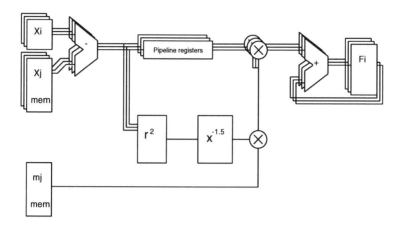

Figure 23   The basic GRAPE pipeline.

The force calculation pipeline evaluates the force from one particle at each clock cycle. Thus, all arithmetic units are working at 100% efficiency during the force calculation. The memory unit provides the position and mass of one particle at each clock cycle. The arithmetic operations necessary to evaluate one interaction are three subtractions, five additions, seven multiplications, one division and one square root operation.

The pipeline could be constructed using floating-point LSI chips. Pipelined floating-point adder/multiplier chips had become available in the mid-80s, but they were still fairly expensive (cost more than 100 000 yen per chip), and the size of the hardware would be quite large, since we would need more than 20 of these chips. At least 15 chips are necessary to perform additions and multiplications, and some additional chips would be necessary to perform the division and square root operations. We thought the system would be too big for us.

We decided to reduce the word length from the standard IEEE floating-point format to something shorter. The actual work of constructing the hardware is much simpler for a short word length than for a long word length. In particular, if the word length is sufficiently short, we can use a ROM (Read Only Memory) chip to perform arithmetic operations. For example, a ROM chip with 16-bit address and 8-bit data (512 kbits in total) can produce any function of two 8-bit numbers in 8-bit width. Such a chip had become available in the mid-80s. The word length of 8 bits has another advantage in that it minimizes the number of IC chips. Standard IC chips typically have 8-bit data width. Thus, a word length longer than 8 bits would increase the number of IC chips nencessary quite significantly. On the other hand, reducing the word length to less than 8 bits does not reduce the number of IC chips.

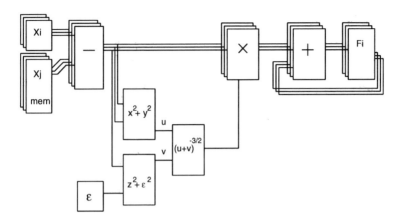

Figure 24   The original plan for the GRAPE-1 pipeline.

The simplest hardware would express the number in 8 bits at all places in the hardware. We first started to design something like that. The position data was expressed in 8-bit fixed point numbers. After subtraction, the result is converted to a logarithmic format, which was used to calculate the force. This conversion is necessary to limit the number of bits in the intermediate results like x^2 and $1/r^3$. If we had done everything in fixed precision, we would have needed 24 bits or more for some of the intermediate results. Use of the logarithmic format for intermediate results makes it possible to use the same 8-bit data length for all intermediate operations without losing accuracy.

The force was again converted to a fixed-point format, and the accumulation is performed in this format. The pipeline would look like that shown in Figure 24. In this figure, all data paths have an 8-bit width. Note that the number of functional units in this figure is only 12, since the ROM tables are used to implement composite operations such as $x^2 + y^2$ or $(u + v)^{-1.5}$. More importantly, all functional units are on a single ROM chip, which cost around 5000 JYE in 1989.

As stated earlier, GRAPE-1 was originally not intended to be useful for real calculation. However, if it could be used for real calculation without too many modifications, it would be very nice. So we did some analysis of the error of the force due to the low accuracy [MIE90].

The result of the error analysis was rather striking. The word length of 8 bits is actually good enough for many simulations, if the position data is expressed with a higher accuracy (e.g., 16 bits).

The r.m.s. relative error of the force between two particles is expressed roughly as

$$< (\Delta F/F)^2 >^{1/2} = [C_1(\epsilon_i/r_{ij})^2 + C_2\epsilon_f^2]^{1/2}, \qquad (4.2)$$

where C_1 and C_2 are the constant of order unity, ϵ_i is the resolution of the fixed number format used for representation of the position, and ϵ_f is the resolution of the floating-point format used to express the intermediate results. The error in the force has two components: the first comes from the error in the position representation; the second comes from the calculations after conversion to the logarithmic format.

The important question is the size of the error which we can accept without sacrificing the validity of the result of the simulations. The answer, of course, depends upon the problem to be solved. In one limiting case, where we are interested in the evolution of the system in the dynamical timescale, one can accept an error comparable to the force itself, as long as the error is not correlated in either time or space.

As an example, consider the simulation of the merging of two galaxies, which takes place in the timescale of 10^9 years. Real galaxies consist of $> 10^{10}$ stars, but we model them with $< 10^6$ stars. In this case, the dominant term of the numerical error comes from the Poisson noise in the potential field, caused by the fact that we model it with a number of particles much smaller than the real number of stars in galaxies. Very roughly speaking, the total force on a particle contains an error $O(N^{-1/2})$. If the force from one particle contains an error of, say, 10%, the error in the total force due to this individual force error is $N^{-1/2}$ times this 10%. Thus, if the individual force error is smaller than 100%, its contribution to the total force error is smaller than the Poisson noise, if the individual force error is unbiased and random.

In practice, we can even accept a numerical error which has a finite correlation time because the Poisson noise of the potential does have time correlation. Consider the orbit of one particle. The force from nearby particles changes quickly, and therefore has a short correlation time. The force from distant particles changes slowly, and therefore has a long correlation time. When we evaluate the instantaneous error of the force on a particle, the force from near neighbors has the largest contribution [MIE90],[Mer96]. However, if we evaluate the cumulative effect integrated over time, contributions from the neighbors and distant particles are equally important [MIE90],[HHM93]. Thus, for numerical errors, we can also accept a longer time correlation for the forces from distant particles.

The theoretical argument in the previous paragraph suggested that we could achieve sufficient accuracy of the final result if $\epsilon_f < 1$. In other words, 8 bits is more than enough. Of course, we need 1 bit for the sign and 3–4 bits for the exponent (the integral part of logarithm). Even so, we can still use 3–4 bits for the fractional part of the logarithm, which offers enough accuracy.

Numerical experiments with an accuracy of 3–4 bits gave acceptable results. This was, for us, actually quite surprising, even though the theory we described above explained the result perfectly.

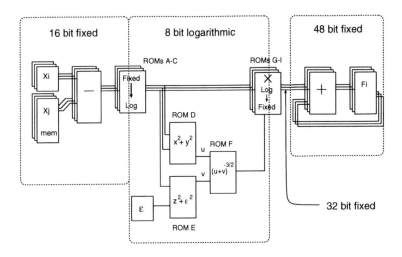

Figure 25   The actual GRAPE-1 pipeline.

The first term of Equation 4.2 must be reasonably small for the system to be actually useful. Fortunately, ϵ_i can be reduced without significantly increasing the total complexity of the hardware. The 512-kbit ROM which we chose to implement binary operations with an 8-bit data length can also implement unary operation for a 16-bit word length. Thus, it can implement the table to convert a 16-bit fixed-point number to 8-bit logarithmic format. Adder LSI chips for 16-bit fixed-point data were available from several different companies.

Figure 25 shows the final version of the pipeline used for GRAPE-1. The position data are expressed in 16-bit fixed-point format. The positions of particles which exert the forces [we call them j-particles, because they are denoted with subscripts j in Equation (4.1)] are stored in three memory units. The coordinates of the particle on which the force is exerted (i-particle) are stored in the input registers of 16-bit adder chips. After subtraction, the results are converted to the 8-bit logarithmic format with a 1-bit sign using a ROM table. The base is two, and three bits are used to represent the fraction.

ROM D performs the calculation $x^2 + y^2$ in logarithmic format. Here, the result is a positive number. So we used five bits for the integral part and three bits for the fractional part. ROM E is the same as ROM D. It performs $z^2 + \epsilon^2$. ROM F performs the addition of two positive logarithmic numbers.

ROM units G–I calculate the force components, such as x/r_s^3, from two inputs x and $r_s^2 = r^2 + \epsilon^2$. These units, at the same time, convert the results to 32-bit fixed-point format. A ROM unit with a 32-bit output consists of two 1 Mbit (64k 16-bit words) ROM chips. The results are accumulated in three

48-bit accumulators. A 48-bit accumulator consists of three cascaded 16-bit adders.

The cycle time of the pipeline unit was limited by the access time of the ROM chips. Initially, we used ROM chips with an access time of 200 ns. We used a rather conservative 4 MHz clock. In the spring of 1990, we replaced the ROM chips with faster ones (85 ns access time), and increased the clock frequency to 8 MHz. The peak performance of GRAPE-1 operating with an 8 MHz clock was 240 Mflops.

For the interface between the host processor and GRAPE-1 we chose the GPIB (IEEE-488) interface. The estimated communication speed necessary to achieve reasonable performance was of the order of 10 KB/s. We thought this speed would be easy to achieve using the IEEE-488 interface. It had the important advantage that the interface boards to the machines which we had were relatively cheap, and the development of the software for these boards was relatively simple. The price of the interface board was quite important, since the total budget we could use for the development of GRAPE-1 was under $ 10 000.

For the IEEE-488 interface we used the TMS 9914 of Texas Instruments. All control logics were implemented using 74HC and 74AC series MSI chips.

The design of GRAPE-1 was started in April 1989 and was finished by June. The board was completed in August and became operational in September. The size of the board is about 40 × 30 cm. The total number of LSI and IC chips on board was 97.

The calculation flow is shown in Figure 26. At the beginning of the simulation, the host processor sends the constant data such as number of particles N and the softening parameter ϵ. At each timestep, the host processor sends the particle data to be stored in the memory. Then, for each particle, the host sends the position. After GRAPE-1 has received three components of position data, it automatically starts the calculation. When calculation is finished, GRAPE returns the calculated force to the host.

We used the IEEE-488 8-bit parallel interface to connect GRAPE-1 and its host. For the transfer of data from host to GRAPE-1, one data element consists of 5 bytes, three bytes for the address and two bytes for the actual data. For each of the 5-byte data elements, GRAPE-1 decodes the address and stores the data element in a requested location. For the transfer of the data from GRAPE to host, GRAPE simply sends 18 bytes of data (six bytes for each component) in a predefined order. Thus, the host first sends $15N$ bytes of data to store the particles in the memory. After that, the host sends 15 bytes and receives 18 bytes. This operation is performed on all particles in the system.

Since the average length of the data packet is short, the software overhead of the IEEE-488 interface on the host must be reasonably small. This was not a problem when we were using an MS-DOS machine, since on MS-DOS the

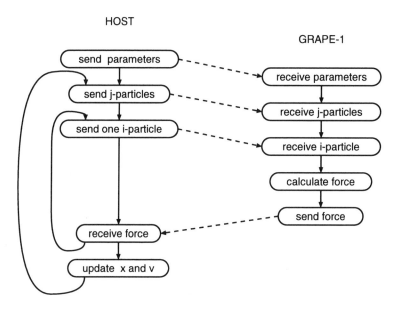

Figure 26   The calculation flow in GRAPE-1.

application program directly accesses the IEEE-488 interface board through a library function. On a UNIX machine, however, this requirement of low latency caused a complex performance problem.

In a multi-tasking operating system such as UNIX, the I/O functions are usually implemented as "kernel" functions which are taken care of by the operating system itself. An application process sends the request messages to the OS through system calls, and the kernel process actually handles the I/O function. This approach causes quite a large software overhead associated with each I/O request. In the case of the UNIX workstation we used as the host of GRAPE-1, one system call took in the order of one millisecond. Since the length of the message we send or receive is around 10 bytes, the communication speed was limited to less than 10KB/s. This speed was quite unsatisfactory. In order to solve this problem, we took two completely different approaches.

The first approach was to insert an intelligent buffer between the host processor and GRAPE-1. The host computer first sends all particles to this buffer. Then this buffer lets GRAPE-1 calculate the force on all particles by looping over all particles and storing the calculated forces in its memory, then sending the forces back to the host in a long message. We used a personal computer as the buffer, and used the IEEE-488 interface to connect between the host and the buffer computer. This method improved the communication speed by a factor of nearly three.

The second approach was to rewrite the interface software for IEEE-488 on the host computer. Around that time, we had two workstations, both from Sony. The older one (Sony NEWS 821) had a classical architecture with a separate processor dedicated to I/O operations. Thus, it was not practical to change the I/O software. However, a newer one (model PWS-1550) did not have an I/O processor, and I/O devices were directly controlled by the CPU. On this machine, the IEEE-488 interface LSI (the TMS-9914 chip from TI, the same chip as we used for GRAPE-1) is connected to the CPU as one of the I/O devices. The control registers of the 9914 chip are directly mapped into the physical address space of the CPU.

Thus, on PWS-1550, the CPU can directly access the IEEE-488 interface chip. In normal I/O operation, however, the register was visible only to the kernel process. Thus, each access to the I/O device still had to go through the system calls. However, it was possible to map the physical registers of the 9914 chip directly into the virtual address space of the application program, by adding new device driver software to the UNIX kernel.

In this approach, the application process has complete control of the communication, and the system call is never used except for the first call to map the 9914 chip. Once the 9914 is mapped into the virtual address space, the application program (to be precise, the low level library function of the GRAPE-1 interface routine) can access the registers of the 9914 chip by the usual memory access operations. In practice, all interface routines are written in the C language, and the registers are accessed through the assignment statements using pointers, just as normal variables in C are accessed.

To send one byte, the interface software first tests the status of 9914. If the status is "not ready", it repeats testing the status register until the status flag changes. Then one byte of data is written to the data register. The read operation is similar.

In this scheme, which is called polling or spinning, the CPU waits for the 9914 to finish the previous command. This approach is usually not used for I/O in a multitasking system, since it wastes the CPU resource, which could otherwise be used by other processes. If we follow the normal procedure, the user process would put itself into a sleep state after writing/reading each byte. The process is woken up from the sleep state by the hardware interrupt generated by the 9914 chip. This approach, however, slows down the communication speed to an unacceptable level because the task switching did take rather a long time. So we decided to use spinning.

This second approach turned out to be quite successful, achieving a speed of about 60 KB/s. With this communication speed, half of the peak performance was achieved for $N = 6 \times 10^3$ [IMES90].

GRAPE-1 was used for the study of merging two elliptical galaxies [OEM91] and the violent relaxation process [FME92a]. Two copies of GRAPE-1 were made by Dr. Sakagami.

The detailed design of the GRAPE-1 hardware was done by Ito. J.M. developed most of the software for GRAPE-1, both the actual simulation program and several variants of the interface libraries.

4.3 GRAPE-2

Though there were many problems for which the reduced accuracy of GRAPE-1 is sufficient, for many other problems a higher accuracy was necessary. In summer 1989, while Mr. Ito was developing GRAPE-1, we started the design of GRAPE-2, which would use the standard IEEE floating-point numbers as the internal representation of data. For all operations except for the first subtraction of the position and the final accumulation of the force, we adopted the IEEE-754 single precision format. For the remaining two operations, IEEE-754 double precision format was adopted.

For GRAPE-2, we obtained a research grant of 2.5 M Yen from the National Astronomical Observatory in summer 1989. This grant covered the total expense of GRAPE-2, including the cost of a VME interface box added to the Sony workstation. We used an SN74ACT8847 LSI chip from TI for 64-bit operations and a 3201/3202 chipset from Analog Devices for 32-bit operations.

The original plan for GRAPE-2 was to develop the hardware for an individual timestep algorithm [Aar63], in which each particle has its own timestep and maintains its own time.

With the individual timestep scheme particle i has its own time t_i and timestep Δt_i. The calculation proceeds in the following steps:

  (a) Select the particle i for which $t_i + \Delta t_i$ is the minimum.

  (b) Obtain the position of all particles at $t = t_i + \Delta t_i$

  (c) Calculate the gravitational force on particle i.

  (d) Update the position, velocity, time, timestep, etc. of particle i.

  (e) Go back to step (a)

For further details see [Aar85] and the references therein. The description of the variable-timestep linear-multistep method similar to the scheme used by Aarseth is given in Krogh [Kro74]. Mann [Man87] gave a comparison of a wide range of time integration algorithms for few-body problems.

The main difficulty of this individual timestep scheme is that, in order to obtain the force on particle i at time $t_i + \Delta t_i$, we need the positions of all other particles in the system at that time. With integration schemes such as Runge–Kutta or extrapolation schemes, we obtain the numerical solution of the differential equation only at discrete points in time. Thus, it is quite difficult to obtain the position of particles at an arbitrary time.

The Predictor-Corrector (PC) scheme can easily provide the numerical solution at intermediate points in time, because it is based on polynomial extrapolation. Of course, the usual implementation of the PC scheme uses a linear combination of the time derivatives with fixed coefficients, which were pre-calculated for the time integration with a constant stepsize. However, in a Krogh-type scheme, the integration scheme is "constructed" at each step, by expanding the fitting polynomial calculated by Newton–Cotes interpolation. Thus, calculating the value of the predictor at an arbitrary time is quite easy.

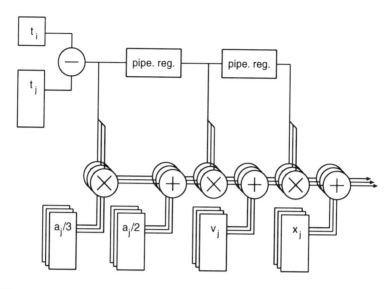

Figure 27   The pipeline to calculate the third order predictor polynomial.

For most astrophysical collisional N-body calculations, the 4th-order predictor-corrector scheme with variable stepsize is used. The positions of all other particles at the time of one particle are calculated using the third order polynomial as

$$x(t + \Delta t) = x + \Delta t v + \frac{\Delta t^2}{2}a + \frac{\Delta t^3}{6}\dot{a}, \tag{4.3}$$

where t is the time of a particle and x, v, a and $\dot{a}$ are the position, velocity, acceleration and the first time derivative of acceleration at time t, respectively. This polynomial is pretty easy to implement in a pipelined hardware, as shown in Figure 27. However, it would significantly increase the total size of the hardware, since the complete pipeline requires as many as 19 arithmetic units (9 multipliers and 10 adders).

Even without this polynomial pipeline, GRAPE-2 would become much bigger and more complicated than GRAPE-1 simply because the word length was four times longer. The addition of the polynomial pipeline made it practically impossible to implement the whole pipeline in a single-board unit. Of course, we could design a multi-board system, but such a system would be difficult to develop and debug. In particular, the frequency of the system clock would be severely limited. A one-board system is by far easier to develop.

To reduce the size of the hardware, we changed the basic design of the pipeline so that it handles x, y and z components sequentially. Figure 28

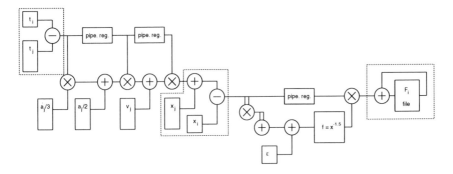

Figure 28   The original plan for the GRAPE-2 pipeline unit. The arithmetic
units and memory units in the dotted box are performed in 64 bit format. All
other operations are performed in 32 bit format.

shows the original plan for the GRAPE-2 hardware. It would fit on board
with the size same as that of GRAPE-1.

For the calculation of $1/r^3$ from r^2, we used the standard technique of a
combination of the range reduction, table lookup and linear interpolation.
Since the IEEE-754 format uses a base-2 floating-point format, we could use
the following relation,

$$f(x \cdot 2^{2e}) = 2^{-3e} f(x), \qquad (4.4)$$

to reduce the range of the argument x to $[1, 4)$. The range of the result is
$(1/8, 1]$.

The straightforward algorithm to evaluate the function $f(x)$ for the range
of x $[1, 4)$ would be as follows. First, we convert x to a fixed-point format.
Then, we perform the table lookup and linear interpolation to obtain the
function f in a fixed-point format. Finally, we convert the result to floating-
point format using a priority encoder and a two-bit logical shifter. In this case,
the exponent of f is calculated by multiplying the even part of the exponent
of x by -1.5. This exponent must be corrected by adding the amount of shift
in the normalization of the fraction of f.

The algorithm we adopted is conceptually more complex than the above
description, but the resulting hardware is actually much simpler. We omitted
the conversion of x to the fixed point format. Instead of the conversion, we
provided separate tables for the ranges of x of $[1, 2)$ and $[2, 4)$. These two
tables can be selected by the lowest bit of the exponent.

Since we provide separate tables of the mantissa of f for x with even and
odd exponents, the exponent of f can have separate tables. As a result, the
range of f is within $(1/4, 1]$ for both the even and odd exponents. Therefore,
the maximum amount of the shift is reduced to one bit.

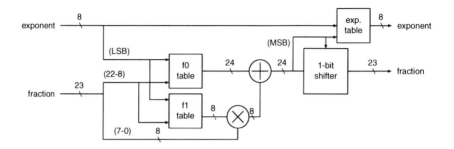

Figure 29   The lookup/interpolation unit of GRAPE-2.

A block diagram of the actual hardware for the interpolation unit is shown in Figure 29. The Least Significant Bit (LSB) of the exponent and upper 15 bits of the fraction are used as the index to obtain the zeroth order term f_0 and the first order term f_1 of the fraction of $f(x)$. The length of f_0 is 24 bits and that of f_1 is 8. The remaining lower eight bits of the fraction are multiplied with f_1 using a multiplier, which is implemented using a lookup table (in the same way as binary operations in GRAPE-1). The result is added to f_0. Note that at this point f is not normalized, since the range of f is $(1/4, 1/\sqrt{2}]$ for the case of odd exponents, and $(1/2\sqrt{2}, 1]$ for the case of even exponents. Thus, f must be normalized by shifting to the right by one bit if the Most Significant Bit (MSB) is 1.

For the exponent, a 512-entry table is prepared. This table accepts all the exponent bits of x and the MSB of the mantissa of f calculated by interpolation (before normalization). This additional input is used for adjustment of the exponent after normalization.

The hardware of the lookup/interpolation unit is quite simple, with one 24-bit adder (24-bit and 8-bit inputs), one 24-bit input 1-bit shifter and three lookup tables. The 1-bit shifter is implemented using 2-input multiplexers. The 24-bit adder is implemented using two 16-bit adders.

We changed the interface to the host from GPIB to VME. This change was necessary to achieve a reasonable data-transfer speed for a relatively small number of particles like $N = 1000$, which was the main target of GRAPE-2. For GRAPE-1, the target number of particles is around 10^4. However, for GRAPE-2 it is smaller, because the main target of GRAPE-2 is long-term evolution of systems in their thermal timescale. GRAPE-1 was intended to be used for simulations of the evolution of systems in their dynamical timescales. GRAPE-2, with its high accuracy, would be applied to simulation of the evolution of the system in a timescale much longer than the dynamical timescale. Therefore, the number of particles that can be used is smaller for simulations using GRAPE-2. The VME bus is a standard 32-bit backplane bus which was used as the I/O bus of many workstations at the time of the design of GRAPE-2. It is still widely used for real-time applications, and the interface adapters between the VME bus and many workstations and personal computers are available from various companies.

The VME interface of GRAPE-2 was designed as a "slave", which is accessed by the host processor through programmed I/O. The VME bus is a system backplane bus which allows multiple masters. Here, a master is the unit which initiates a transaction and issues the address. A slave is the unit which responds to the request of the master. In a write transaction, the master outputs the address and data and the slave receives them. In a read transaction, the master outputs the address and the slave outputs the requested data. The bus protocol is asynchronous, thus a read transaction proceeds as follows.

(a) The master outputs the address to the address lines of the bus.
(b) The master drives the AS (Address Strobe) signal low. This action tells the other units that the address is now valid. The master also drives the DS (Data Strobe) signal low.
(c) Each of the other units decodes the address, and if it falls within the address range assigned to it, it outputs the data of the requested address. After the data lines have settled, it drives the ACK line to low.
(d) The master acquires the data and deasserts the AS/DS lines.
(e) The slave deasserts the ACK line.

The write operation proceeds in a similar way. The only difference is that, with the write transaction, the master drives the data lines and the slave reads in the data on the bus.

The asynchronous nature of the bus allows the co-existence of slow and fast peripherals, since different units on the same bus can operate at different speeds. In other words, the design of the hardware is easy if the requirement for speed is not very high. On the other hand, it is very difficult to achieve high performance, since the sequence of signals needed to transfer one word is rather complex and difficult to be pipelined. In the case of a synchronous bus, all units share a common bus clock. In this case, it is not very difficult to achieve a transfer rate of one word/clock. For the GRAPE-2 system, the data transfer rate we needed was less than 1 MB/s, which is quite easy to achieve with VME.

Most computers which adopt the VME bus as the I/O bus or the system backplane use it in two different methods. In one method, usually called the Programmed I/O (PIO), the VME address space is mapped into the address space of the CPU and the load/store operation of the CPU is translated to the VME bus transaction. In the other method, usually called DMA (Direct Memory Access), part of the physical memory of the system is mapped into the VME address space, and adapter cards access the system memory through the VME bus.

Usually, a PIO operation is used to issue commands to peripherals, while a DMA operation is used to transfer large chunks of data. For example, if a disk interface card is on the VME bus, the CPU sends the commands to disks using a PIO. The disk sends/receives the data using a DMA operation. This is a quite sensible design, since it liberates the CPU from the task of moving data between the main memory and slow peripherals such as a hard disk unit. The DMA operation is therefore particularly useful in a multi-tasking operating system.

On the other hand, a VME interface with DMA capability is more difficult to design compared to an interface which acts only as a slave. Moreover, the software on a host computer using such a board is far more complicated than that used to access a slave VME board, at least in the case of the UNIX operating systems we have used for almost all hosts of GRAPE systems sofar.

As mentioned in the description of GRAPE-1, almost all UNIX-based computers today offer the memory mapping function, which maps the address space of the I/O bus directly into the address space of the CPU. In a UNIX operating system, this mapping can be done into the virtual address space of a user process. Thus, once this mapping is accomplished, the application program can access the board on the I/O bus as part of its memory space. In the C language this space can be accessed simply through pointers. Thus, even though we deal directly with the hardware, we can use the same source program for the interface library for almost all host computers.

In the case of DMA, the access is more complicated. The standard method of data transfer using DMA in a UNIX operating system can be described as follows (in the case of a read operation):

(a) The user process allocates the target memory area and sends the data transfer request to the kernel using a read/write system call. The user process itself is put into the sleep state.

(b) The kernel process allocates an area in the main memory to be used as the buffer.

(c) The kernel process issues the command to the device, requesting the data. This command should specify the length of the data and the starting address of the buffer memory. The kernel process also goes into a sleep state.

(d) The device actually transfers the data. When the data transfer is complete, it sends the hardware interrupt signal to the CPU.

(e) The interrupt handler in the kernel wakes up the kernel process.

(f) The kernel process transfers the data in its buffer to the target memory area of the user process. After all data has been transferred, the user process is woken up.

This rather complicated procedure naturally requires a fairly large overhead per transaction, which was of the order of one millisecond on workstations of the late 1980s. Thus, for the same reason as the system-provided IEEE-488 interface being unusable for GRAPE-1, a standard implementation of DMA transfer on VME was unacceptable for GRAPE-2.

In principle, we could eliminate the software overhead by allowing the user process to issue the command directly to GRAPE, and GRAPE hardware to access the user memory space directly. However, to develop such software is rather complicated, since it requires an in-depth understanding of the virtual memory system. In addition, such software is difficult to port to other host computers.

For GRAPE-2, the data transfer speed of the PIO operation was sufficient to achieve a reasonable efficiency, so we decided not to use the DMA operation. Up to now (1996), the only GRAPE machine which uses the DMA operation is the Teraflops GRAPE-4. All other machines use the slave VME slave interface, which is essentially the same as that of GRAPE-2.

The debugging of GRAPE-2 took quite a bit more time than that for GRAPE-1. There are many reasons for this, but the most critical is that we did not have enough knowledge of using ultra-fast CMOS VLSI chips.

The force-calculation part of the GRAPE-2 hardware was operational by the spring of 1990. At that time, we made a decision to abandon the predictor pipeline.

One of the reasons why we abandoned the predictor pipeline is that we could achieve a reasonable performance without the predictor pipeline by slightly changing the basic algorithm. In the original implementation of the individual timestep algorithm, particles are allowed to have an arbitrary timestep and time. Thus, different particles have different times, and only one particle can be integrated at a time. However, if we force the timesteps to, say, integer powers of two, many particles share the same time and timestep. Thus, the calculation cost of the predictor is significantly reduced, since the predicted positions can be used to evaluate the forces on all particles that share the current time.

This algorithm was first used by McMillan [McM86] to increase the vector length for the force calculation. See Makino [Mak91a] for details of the implementation and performance on GRAPE-2.

GRAPE-2 was used for a number of problems, including the merging of galaxies with central black holes [EMO91] and the evolution of early solar systems [IM92a], [IM92b].

4.4   GRAPE-1A and GRAPE-2A

GRAPE-1A [FIM$^+$91] is an improved version of GRAPE-1 developed by
Toshiyuki Fukushige as his undergraduate thesis project. The differences from
GRAPE-1 are summarized as follows:

- The interface to the host is changed to VME.

- The pipeline is changed to handle particles with non-equal masses.

- The effective word length is increased by one bit.

- The hardware to construct the list of neighboring particles is added.

- The hardware to calculate the potential energy is added.

The first two changes were introduced to use the $O(N \log N)$ algorithm,
to be described in Section 5.2. The improvement in accuracy is rather experimental, to test the use of fixed-point adders to perform multiplications in
logarithmic format.

The list of neighbors is used to combine GRAPE with the hydrodynamical
simulation with SPH (Smoothed Particle Hydrodynamics) [UFM$^+$93].

Two copies of GRAPE-1A were built. One was used at the National Observatory of Japan, and the other was used at the Department of Astronomy,
University of Tokyo.

GRAPE-2A [IMF$^+$93] was our first trial in using a GRAPE-type architecture for N-body problems outside astrophysics. It was developed in joint
research with the central research laboratory of Taisho Pharmacy. The main
target of GRAPE-2A is Molecular Dynamics (MD) simulations of protein
molecules.

The most important difference between MD simulations and gravitational
N-body simulations is the force law. In gravitational N-body simulation, the
force law is pure $1/r^2$ Newtonian gravity. On the other hand, in MD simulations, the force law is more complex. If the atoms are charged, they interact
through the $1/r$ Coulomb potential. However, atoms also interact through
the van der Waals force, for which many different models exist. Therefore, the
force needs to be programmable.

To implement a programmable force law is not very difficult, using the same
table lookup and linear interpolation as is used in GRAPE-2. However, the
range reduction technique cannot be applied, since the simple relation such
as Equation (4.4) cannot be used unless the force law is expressed as $f \propto r^\alpha$.
Thus, we have to supply one table for each value of the exponent. In other
words, the size of the table is much larger, since all bits of the exponent must
be used as an index. In the case of the gravitational force, only the LSB need

to be used. Thus, the size of the table would be $2^7 = 128$ times larger if we have to provide the table for the entire range of r^2.

Fortunately, for almost all applications the actual size of the table necessary is much smaller, since the effective range of the force is usually small. Thus, the force needs to be defined only for a relatively small range of distances. Consider the case of the Lennard–Jones potential, which is expressed as

$$\phi_{LJ} = -r^{-6} + r^{-12}. \tag{4.5}$$

It has the minimum value at the distance r_0, and increases very rapidly for $r \to 0$. Thus, two atoms would not approach a distance smaller than, say, $r_0/3$. For the limit of $r \to \infty$, the force converges to zero as r^{-7}. Thus, the force from a particle at the distance $10r_0$ is smaller than the force from a particle at a distance around r_0 by a factor of 10^{-7}, and is practically negligible.

The function evaluator of GRAPE-2A is designed so that it can calculate both the Coulomb force ($1/r$ force) and arbitrary forces. Figure 30 shows the structure of the function evaluator. The main difference from the interpolation unit of GRAPE-2 is that the floating-point units are used to perform linear interpolation. This change is necessary to evaluate the force which crosses zero at a finite distance. Also, the use of the floating-point arithmetic units simplified both the design and error analysis of the interpolation unit.

The tables for f_0 and f_1 now simply output the full 32-bit floating-point numbers. These tables have 2^{15} entries for one function, and four functions can be stored in the table. The tables are implemented using 1 Mbit SRAM chips.

The interpolator has two different operation modes. For the case of the gravitational force, only one bit of the exponent is supplied to the table. This is achieved by left-shifting the input of the shifter by three bits. In this case, the linear interpolation unit gives the floating point value of $f(x) = x^{-1.5}$. The exponents are fed to the exponent table, which gives the offset of the exponent. This offset is added to the exponent part of the output of the linear interpolator in the exponent ALU.

In the case of arbitrary functions, the lower four bits of the exponent are fed to the table. Since the table has 2^{15} entries, the higher 11 bits of the mantissa are also fed to the table. This gives roughly single precision accuracy for the force after linear interpolation. In this mode, the exponent table stores the 1-bit information, which says whether the function is non-zero for that exponent or not. Since four bits are supplied to the interpolator, it can evaluate the function for the range of $x_0 \le x < 2^{16}x_0$. The argument x is actually r^2. The range of r is therefore $r_0 \le r < 256r_0$, which is more than enough, as we stated earlier.

For GRAPE-2A, we returned to the basic fully parallel pipeline design from the serial pipeline of GRAPE-2. The clock speed was increased to 6 MHz. These changes made GRAPE-2A 4.5 times faster than GRAPE-2.

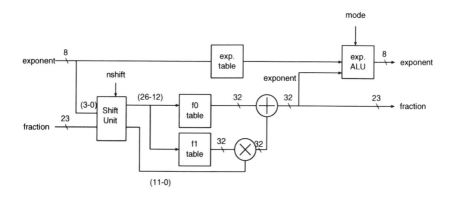

Figure 30   The function interpolation unit of GRAPE-2A.

The GRAPE-2A board has a number of additional features which are necessary to make it useful for MD calculation of complex organic molecules such as proteins. In the following, we briefly describe each of these additional features.

First, GRAPE-2A can calculate the forces from a subset of particles stored in the memory, using indirect addressing (Figure 31). This capability is necessary to implement algorithms like the linked-list algorithm discussed in Section 3.4. The indirect addressing unit is essentially an SRAM which stores the indices of particles. During the force calculation, the counter supplies the consecutive addresses to the memory, which generates the actual address supplied to the memory unit.

In the actual calculation using the linked-list algorithm, first the host computer sends the data of all particles in the system. Then, for each cell, the host sends the indices of all particles in its 27 neighboring cells (including itself), and calculates the force from these particles to the particles in the cell. In this way, the total amount of communication is reduced quite significantly. Instead of sending three position coordinates and charge (28 bytes), we need to send just one index (2 bytes).

Second, the van der Waals force used in the simulation of proteins has the functional form

$$\phi_{\mathrm{vdw}} = a_{ij} f(r_{ij}/b_{ij}), \qquad (4.6)$$

where a_{ij} and b_{ij} are the coefficients which depend upon the species of the two atoms, r_{ij} is the distance between two atoms, and $f(x)$ is a given function. The number of species is relatively large (more than 50), because the van der Waals interaction depends upon the location of the atoms in molecules. Therefore, we had to implement rather complex table lookup hardware for this part, which is shown in Figure 32.

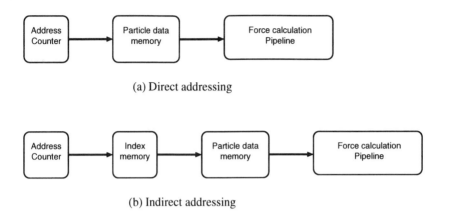

(a) Direct addressing

(b) Indirect addressing

Figure 31   The hardware for indirect addressing. (a) Direct, consecutive addressing, and (b) Indirect addressing.

In the case of the calculation of the Coulomb or gravitational forces, the table lookup hardware is bypassed. The multiplier to calculate r_{ij}/b_{ij} simply passes through r_{ij}, and the particle index is directly fed to a_{ij} memory.

Two additional copies of GRAPE-2A were build and installed in ETL and at a research project taking place under the Science and Technology Agency.

The development of GRAPE-2A was started in summer 1991. It became fully operational in May 1992.

4.5   WINE-1

WINE-1 [FMI+93] is special-purpose hardware designed to perform the Ewald method, which is a standard technique to evaluate gravitational or Coulomb interaction in periodic boundary. The basic idea of the Ewald method is to separate the interaction into two terms, one with a long wavelength and the other with a short wavelength. A filter is used to separate long- and short-wavelength components. The short-wavelength component is evaluated by direct summation, while the long-wavelength component is evaluated by Discrete Fourier Transform (DFT). In order to achieve a fast convergence in both the short- and long-wavelength calculations, a Gaussian filter is used to separate two components.

Calculation of the short-wavelength contribution is the same as that for the usual gravity or Coulomb calculation, except that the functional form is different [Ewa21], [Kit86]. GRAPE-2A, with its programmable function table, can handle the short-wavelength, real-space part.

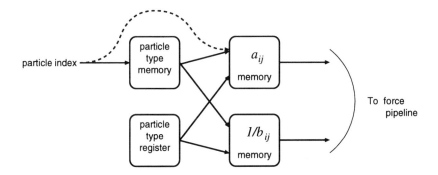

Figure 32   The hardware for the coefficients table for van-der-Waals force.

To evaluate the long-wavelength contribution, we (1) calculate the Fourier series expansion, (2) apply the Gaussian filter in the wavenumber space, solve the Poisson equation, and (3) calculate the potential or force at the location of each particle in the system, by adding the contributions of all wavenumbers. The calculation costs of the first and third parts are $O(Nm)$, where N is the number of particles and m is the number of Fourier components. The cost of the second part is $O(m)$.

WINE-1 (Wave space INtegrator for Ewald method) is the hardware to perform the $O(Nm)$ part of the calculation described above [FMI+93]. The first part of calculation is expressed as

$$
\begin{aligned}
b_{\mathrm{s},n} &= \sum_{j=1}^{N} m_j \sin(2\pi \vec{k}_n \cdot \vec{r}_j), \\
b_{\mathrm{c},n} &= \sum_{j=1}^{N} m_j \cos(2\pi \vec{k}_n \cdot \vec{r}_j).
\end{aligned} \tag{4.7}
$$

The structure of the pipeline is shown in Figure 33. It calculates the inner product of the position and wave number vector. The result is supplied to a unit which calculates trigonometric functions, and the result is accumulated.

The second part of calculation is expressed as

$$
\begin{aligned}
\vec{c}_{\mathrm{s},n} &= b_{\mathrm{s},n} a_n \vec{k}_n, \\
\vec{c}_{\mathrm{c},n} &= b_{\mathrm{c},n} a_n \vec{k}_n,
\end{aligned} \tag{4.8}
$$

where

$$
a_n = \frac{\exp(-\pi^2 \eta^2 k_n^2)}{k_n^2}. \tag{4.9}
$$

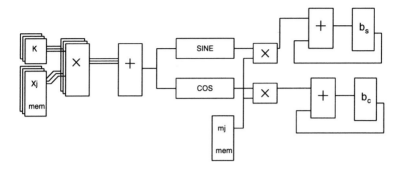

Figure 33   The pipeline for the first part of the Ewald method.

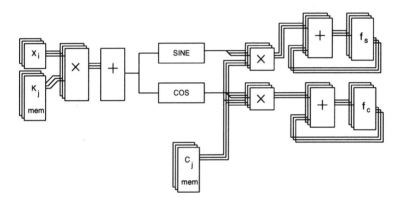

Figure 34   The pipeline for the third part of the Ewald method.

These calculations are performed on the host computer.

Finally, the third part is expressed as

$$\vec{f}_{\mathrm{s},i} = \sum_{n=1}^{n_k} \vec{c}_{\mathrm{c},n} \sin(2\pi \vec{k}_n \cdot \vec{r}_i),$$

$$\vec{f}_{\mathrm{c},i} = \sum_{n=1}^{n_k} \vec{c}_{\mathrm{s},n} \cos(2\pi \vec{k}_n \cdot \vec{r}_i), \tag{4.10}$$

$$\vec{F}_{\mathrm{wv}}(\vec{r}_i) = -\frac{2m_i}{L^3}(\vec{f}_{\mathrm{s},i} - \vec{f}_{\mathrm{c},i}), \tag{4.11}$$

and the pipeline for the third part would look like that shown in Figure 34.

If we compare Figures 33 and 34, we can see that the pipeline for the third part can be used for the first part as well. Therefore, we only need to build one pipeline.

WINE-1 was our first trial to implement the Ewald method in hardware. To further simplify the hardware, we implemented only three accumulators, which are used to accumulate either the sine or cosine part, instead of six accumulators.

We assembled WINE-1 from off-the-shelf components, similar to those used in GRAPE-1/1A. For all calculations, the number format is 16-bit fixed-point. Evaluation of the trigonometric functions is implemented as a table lookup.

WINE-1 operates on 12 MHz clock at a speed equivalent to 240 Mflops.

4.6   GRAPE-3

In the fiscal year of 1990, we got two small research grants, one from the Ministry of Education (MoE), the other from the Yamada Foundation for Promotion of Science. The combined amount of grant was 10 M Yen. The original plan for these grants was to build multiple copies of systems like GRAPE-1A to achieve higher speeds.

Of course, we had realized that a single pipeline of GRAPE-1A could be easily implemented in a single custom LSI with the LSI manufacturing technology at that time. The pipeline of GRAPE-2 could fit in a chip too. However, according to quotes we got from several LSI foundries, 10 M Yen was not quite sufficient for a custom LSI.

In the U.S.A., MOSIS had been offering the service of manufacturing experimental LSI chips in small quantities for a very low cost. Such an organization is only now being formed in Japan. In addition, most Japanese LSI foundries are specialized for commercial products which are manufactured in large volume. Thus, they were (are) not willing to get a contract to develop chips which will be manufactured in only tiny quantities, like a few hundred.

We spent quite a bit of time negotiating with several manufacturers, without much success. Finally, the Electronic Laboratory of Fuji-Xerox Corp. agreed to develop the chip as a joint research project. We got to know them through the Nissei-Sangyo Corp., which was selling ASIC design software from Silicon Compiler Systems. Fuji-Xerox Laboratory was one of the customers of this system.

We decided to develop a relatively simple chip, essentially an extension of the pipeline in the GRAPE-1A system. It was not possible to implement the pipeline of GRAPE-1A directly into a chip. The main problem is the approach of using large lookup tables to implement arithmetic operations. The ROM tables on GRAPE-1/GRAPE-1A consumed more than 10 million transistors, which was impossible to integrate on a single chip in 1990. Practically speaking, the number of transistors we could use is around 2×10^5.

To reduce the transistor count, we completely changed the design of the pipeline for GRAPE-3 so that it used the usual arithmetic logic units for calculations. However, to minimize the design time, we decided to use a logarithmic format similar to that used in GRAPE-1A.

The main advantage of using the logarithmic format is, in this case, that multiplication and division are reduced to addition and subtraction. Moreover, the square operation is reduced to one-bit shift, which requires no hardware. The evaluation of $f(x) = x^{-1.5}$ is also reduced to one addition. This all makes the design of the pipeline and the analysis of the error quite a bit simpler than that for standard floating-point representation. The other advantage of the logarithmic format was that we had done pretty comprehensive analysis of the error for this format [MIE90], so it was relatively easy to evaluate the effect of the change in the accuracy.

On the other hand, the addition and the conversion between the logarithmic format and the fixed point format are quite complex. The main challenge here is that relatively little was known about the implementation of logarithmic arithmetic. A few papers were available on the subject [KR71], [SA75], [Kor93], but they did not discuss the cost of implementation and accuracy achieved in sufficient detail. Therefore, we had to figure out what was necessary by ourselves.

Figure 35 shows the circuit for fixed-to-log conversion. In the first step, the 2's complement format is changed to the sign+magnitude format. Then, the leading zeros in the absolute value are counted by the priority encoder. The output of this priority encoder is used as the "integer" part of the logarithm. The absolute value is then left-shifted using the logical shifter. Finally, the upper six bits of the output of the shifter are supplied to the table, which converts the fractional part to logarithm. This table has a size of 128 5-bit words. In actual hardware, this table is implemented as a PLA (Programmable Logic Array), not as a ROM, because PLA implementation turned out to be smaller. Thus, this part is essentially the same as the hardware used to convert the fixed-point format to floating-point format. The only difference is the final lookup table. The size of the table is practically negligible.

The design of the actual hardware needs some more consideration. In the GRAPE chip, the input positions are represented in 19-bit fixed-point format, and the result of the subtraction $x_i - x_j$ becomes 20-bit (19-bit for absolute value and 1-bit for sign). The conversion unit from fixed to float converts it to the 13-bit logarithmic format, with 1 sign bit, 7 bits for the integer part of the base-2 the logarithm, and the remaining 5 bits for the fractional part. For position, exponents need not be more than 5 bit. However, here we used a 7-bit exponent, since for $m_j r_{ij}^{-3}$ we need more than six bits. Having a 7-bit exponent here at least simplifies the consideration of overflow.

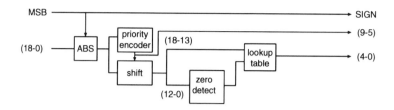

Figure 35   The circuit to convert the fixed-point number to a logarithmic number in GRAPE-3.

There are several additional logic steps necessary to obtain correctly rounded results. We decided to implement unbiased rounding in a way similar to that defined in the IEEE-754 standard. In hindsight, it would be much simpler to have implemented the round-to-zero, and the achieved relative accuracy would have been practically the same. In the case of the general-purpose floating-point unit, it is important that the rounding is unbiased, since there are cases where the biased rounding degrades the accuracy of the final result drastically. Consider the case in which we add many small numbers to one big number. If the rounding is biased, the error accumulates linearly. However, if the rounding is unbiased, the error accumulates only as the square root of the number of terms added, if the terms are random. (Of course, if the terms are exactly the same, the "unbiased" rounding still causes linear growth of the error.) A general-purpose floating-point unit should be able to give the best result for any conceivable use of it, because it is general-purpose. However, in the case of an arithmetic unit used in a specific location of a pipeline to calculate some physical quantity, we can optimize its characteristics to its special requirements.

In the case of fixed-to-float conversion after the subtraction, we can allow a biased rounding, since the bias can be corrected by software on the host by a single multiplication. The bias here means that the result of the conversion, on average, is b times the unbiased result. Thus, the calculated force would be affected by the factor $1/b^2$. Since the forces from all particles suffer (on average) the same bias, the bias on the final result can be corrected by dividing the calculated result by the bias factor.

The logarithmic format cannot express zero. We used one special bit to indicate that the number is zero. In the case of the general-purpose floating-point units, the value zero is usually implemented as a special bit pattern. With this approach, we can use all bits of the word to express a number. If we use one bit to express the value zero, either the exponent or the mantissa would be cut by one bit. If the number of bits in one word is strictly limited to, say, 32- or 64-bit, the loss of one bit is critical. It makes sense to have

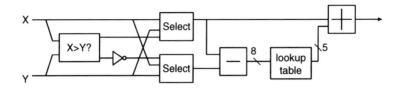

Figure 36   The unsigned logarithmic adder in GRAPE-3.

some additional circuit to detect/set zero. However, for number representation within the hardwired pipeline, to have the zero-flag is the better solution, since the circuit to detect/set zero is replaced by a single wire to propagate the zero flag.

The addition of two logarithmic numbers is calculated using the following relation:

$$\log(X + Y) = \log X + \log(1 + Y/X) = \log X + \log[1 + \exp(\log Y - \log X)]. \tag{4.12}$$

In an algorithmic form, it would be expressed as

```
if (x < y) {
    tmp = x; x = y; y = tmp;
}
result = x + table[x-y];
```

The hardware is fairly simple. Figure 36 shows the unsigned logarithmic adder unit.

Here, we take advantage of the fact that both X and Y are positive numbers. If $X > Y$, the second term has the range $[0, 1)$. This narrow range simplifies consideration of the accuracy, and makes it possible to use a simple and small circuit.

If we regard the numbers in the fixed-point logarithmic format as the integer numbers, the relation between the original number X and the number in the logarithmic format x is expressed as

$$X = 2^{x/32}, \tag{4.13}$$

or

$$x = 32 \log_2 X. \tag{4.14}$$

The range of x^2 is $[1, 2^{40})$. That of $r^2 = x^2 + y^2 + z^2$ is, in theory, $[1, 2^{42})$. Thus, at first sight, one might think that the table needs 41×32 entries. However, in practice we do not have to supply the table if $Y < X/64$, which is $x - y > 210$, since the result, even after being correctly rounded, is the same as x. Thus, the actual hardware needs to have the logic to suppress the addition if $x - y > 210$. To simplify the hardware, we chose to supply the table for $x - y < 256$.

Here, again, we decided to implement unbiased rounding. For addition, unbiased implementation is preferred over a biased implementation, since biased addition introduces an error which depends upon the orientation of the relative position vector, which cannot be corrected by software.

Note that, for the majority of cases, the actual output of the table is either 0, 1 or 2. Thus, it looks rather wasteful to use ROM for this table, since almost all the bits of the table are 0. However, we did not make any further effort to reduce the table size, since the table was already small enough.

It would be worth discussing several possible ways in which to reduce the table size, since they can be used to implement the logarithmic arithmetic with an accuracy higher than that used in GRAPE-3.

The most straightforward approach is to use the piecewise polynomial interpolation to approximate the function, as we did for $f(x) = x^{-1.5}$ in GRAPE-2/2A. This approach can be applied to any reasonably smooth function.

There is a very different approach that takes advantage of the characteristic of the function to be evaluated. The function we evaluate here is

$$s(x) = \log_2(1 + 2^{-x}), \tag{4.15}$$

where x is a positive real number. The range of x is $[0, -\log_2 \varepsilon)$, where ε is the machine epsilon (here we ignore the additional table necessary to implement correct rounding). The calculated value of $s(x)$ needs to have the absolute error comparable to ε or smaller.

Consider the case in which we use p bits for the fractional part of the base-2 logarithm. The relative accuracy of the numbers (in real value) is $\ln 2 \cdot 2^{-p}$. Thus, the range of the argument x is $[0, p)$. The range of function $s(x)$ is $(0, 1]$.

The function $s(x)$ has two important properties. First, it converges to $\log_2 2^{-x}$ for large x. Second, as a result, it's first derivative converges to zero for large x. As a result of the first property, we can use the approximate relation $s(x + 1) = s(x)/2$ for $x \geq p/2$. This technique cuts the size of the table in half.

The second property further reduces the table size (see Koren [Kor93] and references therein). The point here is that the resolution of the table can be coarser for larger values of x. For $1 < x < 2$, $s(x)$ changes from 0 to 0.585. The range of $s(x)$ is smaller for larger x, and yet the accuracy we need for s is the absolute accuracy. Since the relation $s(x) \propto 2^{-x}$ holds approximately, all derivatives of $s(x)$ are also proportional to 2^{-x}. Thus, if we use an r-th order interpolation formula, the error is proportional to $2^{-x}\Delta x^{r+1}$, where Δx is the interval of the table. Thus, to keep the interpolation error constant over all ranges of x, we can increase Δx in proportion to $2^{x/(r+1)}$. This is particularly advantageous in the case of the zeroth-order polynomial (table lookup with no interpolation), since the size of the table is reduced from $p \cdot 2^p$ to 2^{p+1}. The first 2^p entries store the table for $[0, 1)$. The next 2^{p-1} entries store the table for $[1, 2)$. In general, we use only 2^{p-i} entries for the table for $[i, i+1)$. This is implemented by a shifter and a small lookup table.

If the order of the interpolation polynomial is higher, a similar technique can still be used. Roughly speaking, the total number of entries is given by $(2 + r)2^{p/(r+1)}$. Note that, with this technique, it is perfectly feasible to implement a logarithmic adder of 24-bit relative accuracy (better than IEEE-754 single-precision arithmetic) with the total area *smaller* than that for an IEEE-754 single-precision floating-point multiplier, if we use $r = 2$.

Conversion from the logarithmic format back to the fixed-point format is similar to the conversion of the floating-point format to fixed-point format, except that we first convert the fractional part of the logarithm to the fractional part of the floating-point format. Then the fractional part is shifted according to the value of the exponent. The sign bit is sent directly to the ALU to control its operation mode (add/subtract).

The overall structure of the pipeline is similar to that of GRAPE-1A. The calculation of $r^3 = (r^2)^{1.5}$ is implemented as one addition, and the division m/r^3 as one subtraction.

Figure 37 shows the mask pattern of the chip. The size is roughly 8mm × 8mm. The chip was designed using the Genesil system of Silicon Compiler Technology and fabricated by National Semiconductor using their $1\mu m$ process. The total number of transistors is 1.08×10^5. Three blocks in the left-top region handle position subtraction and conversion to logarithmic format. Four big blocks at the bottom handle conversion of the calculated force and potential from logarithic format to fixed-point format and accumulation. Each of these four blocks comprises one shifter, one adder and one register.

The four shifters used to convert from the floating-point to the fixed-point format occupy a significant area. This is partly because of the design method we used. The Genesil system generates the transistor-level layout directly from the logic symbols, such as an adder, a multiplier and a shifter. This approach is quite different from gate arrays and the cell-based design now widely used for small-quantity LSI designs.

The main advantage of the Genesil-type approach is that the die area can be smaller than a gate-array design, because the layout of transistors and the wiring for each functional unit is highly optimized. For example, transistors of different sizes can be used for different functional units. Also, the orientations of transistors can be changed.

Even within one functional unit, one can change the size of the transistor. For example, the transistors which output the data to the outside world need to be large in order to drive signal lines which can be rather long. On the other hand, for transistors whose output signals are used only internally, we could use small transistors.

$\overleftarrow{2\ \mathrm{mm}}\!\!\!\!\overrightarrow{}$

Figure 37   Mask pattern of the pipeline chip of GRAPE-3.

The disadvantage is that to develop the software to generate complex functional units such as multipliers or shifters seems to be a difficult task for the developers of the CAD software. As a result, the algorithm used in these functional units was relatively simple. In some cases it is optimized for the area but is rather slow, and in other case, the size is unnecessarily large. For example, a parallel multiplier is generated using a simple algorithm which resulted in a propagation delay proportional to the size of the operand n. In principle, one could design a multiplier which can generate the result in $\log n$ time, using the Wallace tree technique.

For the shifter, the algorithm we used required the number of transistors to be proportional to $\max(n, m)^2$, where n and m are the length of the input and output words, respectively. The shifter could be implemented with the number of transistors proportional to $n \log n$.

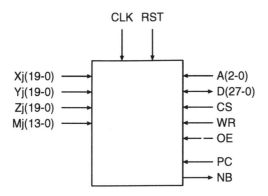

Figure 38   I/O interface of the pipeline chip of GRAPE-3.

If we used the gate array design, the number of transistors would be reduced significantly. However, the average size of the transistor used would have become larger. Therefore, it is not clear whether we could reduce the size of the chip if we have used the gate array design.

One nice thing about the cell compiler design is that the chip looks much prettier than a gate array chip. We can see which part of the chip is used for which function. With the gate array design, the transistors on the silicon are all the same.

Figure 38 shows the I/O interface of the GRAPE chip. During the calculation, it reads three coordinates and the mass at each clock cycle. The 28-bit wide data bus is used to store the position of a particle and softening parameter ϵ, and to output the calculated acceleration and potential.

The first GRAPE chips were packaged in 181-pin ceramic PGA packages. We also manufactured chips in a 208-pin QFP package. The power dissipation was less than 2 W and a QFP package had sufficiently low heat resistance.

As stated earlier, the pipeline chip was developed in collaboration with Fuji Xerox Laboratory. The actual design proceeded as follows. First, we (J.M.) determined the complete specification of the chip, which included the bit-level software simulator of the whole pipeline written in the C language. Then Fuji-Xerox designed the actual chip. In the meantime, we (mainly Sachiko K. Okumura in our lab) designed the board to house the chip. At the time Fuji-Xerox finished the chip design, J.M. prepared the test vectors for the design verification of the chip. The chip designers used the scan path design method to achieve a reasonable testability of the chip. The scan path test vectors, however, can be used only to test the internal logic (we did not use the boundary scan method for this board). So the hand-made (actually generated by several hundred lines of C program) test vectors are crucial for testing.

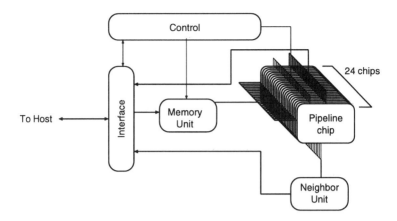

Figure 39   Block diagram of the GRAPE-3 board. (Reproduced from Makino *et al.* (1997).)

The development of the specification started in August 1990. The first draft of the specification was ready by October and the final version by November. The design entry of the chip was completed in December 1990, and tape out in February 1991. Thus, the chip was developed in six months, essentially by three people (J.M. and two designers from Fuji Xerox).

Figure 39 shows the design of the first GRAPE-3 board. The interface to the host is a VME bus, and the board operates as a slave device. Most of the control logics of the board were implemented using small PLDs. We made two GRAPE-3 boards, each of which housed 24 GRAPE chips.

The original design goal was to operate the board with a 20 MHz system clock. If this speed had been achieved, one chip would have offered a peak speed of 600 Mflops and the speed of the total machine would have been 28.8 Gflops. Unfortunately, because of several electrical problems, we could not operate the boards with a clock speed faster than 10 MHz, and the peak speed was reduced to 14.4 Gflops.

The total system became operational in September 1991. The measured best performance of the two-board GRAPE-3 system was 9.9 Gflops, achieved for the direct simulation of a 200 000 particle system.

A few months after completion of GRAPE-3, we started the new development of a version with minor changes. We changed the design in many places to solve the electrical problems and to increase the clock speed. The second design reduced the total number of chips from 24 to a maximum of 16, and only boards with four or eight chips were actually manufactured. This board is called GRAPE. We made six wire-wrapped board with four chips. More than 20 PCB boards with eight chips were manufactured and sold by Computer

Table 5   GRAPE-3A systems installed outside our lab.

Institutes	Configuration	Speed (Gflops)
Hokkaido University	3 boards	14.4
Tohoku University	3 boards	14.4
National Astronomical Observatory	3 boards	14.4
University of Tokyo	2 boards	9.6
Kobe University	1 board	4.8
Edinburgh University	1 board	4.8
Kiel University	1 board	4.8
Max Planck Institute at Heidelberg	1 board	4.8
Max Planck Institute at Garching	1 board	4.8
Marseille Observatory	6 board	26.4
Princeton University	1 board	2.4
University of California at Berkeley	1 board	4.8
University of Arizona	1 board	4.8
Columbia University	1 board	4.8

Garden Ltd. Table 5 lists the institutes which own GRAPE-3A systems as of winter 1996.

The copies of GRAPE-3/3A are used for a number of applications, like the simulation of clusters of galaxies, mergings of galaxies with and without black holes, evolution of small groups of galaxies, the formation of galaxies, the formation of globular clusters, early evolution of globular clusters, and so on. In many institutes, it is currently used as the main computing engine for particle-based astrophysical simulations. Its speed is still far better than that of workstations of a similar price tag. The duplication cost of a single-board GRAPE-3AF with eight chips is around $ 10 000. Its peak speed is 5 Gflops, and half of the theoretical peak is actually achieved for 20K particles or so.

4.7   HARP-1

HARP is an acronym for the Hermite accelerator pipeline. It is an extension of the GRAPE architecture to give hardware support for the Hermite integration scheme [Mak91b], [MA92].

The Hermite integration scheme is based on the Hermite interporation method. Hermite interporation is similar to the Newton–Cotes interpolation, which is the basis of the variable-stepsize linear-multistep integrator used in the Aarseth method.

In Newton interpolation we only use the values of the function f. With the Hermite interporation, we use the values of the derivatives of f in addition to f to construct the interporation formula. In the case of the gravitational N-body system, the first derivative of the acceleration can be calculated for a small additional cost. The acceleration and its first time derivative are given by

$$\mathbf{a}_i = \sum_j Gm_j \frac{\mathbf{r}_{ij}}{(r_{ij}^2 + \epsilon^2)^{3/2}} \tag{4.16}$$

$$\dot{\mathbf{a}}_i = \sum_j Gm_j \left[\frac{\mathbf{v}_{ij}}{(r_{ij}^2 + \epsilon^2)^{3/2}} - \frac{3(\mathbf{v}_{ij} \cdot \mathbf{r}_{ij})\mathbf{r}_{ij}}{(r_{ij}^2 + \epsilon^2)^{5/2}} \right], \tag{4.17}$$

where

$$\mathbf{r}_{ij} = \mathbf{x}_j - \mathbf{x}_i, \tag{4.18}$$

$$\mathbf{v}_{ij} = \mathbf{v}_j - \mathbf{v}_i. \tag{4.19}$$

Here, ϵ is the softening parameter.

The calculation of the acceleration requires nine additions, eight multiplications, one division and one square root. The calculation of the first time derivative (we hereafter call it "jerk") requires another 13 multiplications and 11 additions. On most modern computers, square root and division are more costly than addition or multiplication by a factor of 10 or more. So the additional cost of calculating jerk is typically less than 50%.

With the jerk calculated directly, the construction of the higher order integrator is simplified significantly. For example, the simplest explicit scheme is now second order in time, instead of first order in time. In the following, we present the complete formula for a 2-step, 4th order predictor-corrector scheme. The predictor is given by:

$$\mathbf{x}_{\mathrm{p}} = \frac{\Delta t^3}{6}\dot{\mathbf{a}}_0 + \frac{\Delta t^2}{2}\mathbf{a}_0 + \Delta t \mathbf{v}_0 + \mathbf{x}_0 \tag{4.20}$$

$$\mathbf{v}_{\mathrm{p}} = \frac{\Delta t^2}{2}\dot{\mathbf{a}}_0 + \Delta t \mathbf{a}_0 + \mathbf{v}_0, \tag{4.21}$$

where $\mathbf{x}_{\mathrm{p}}$ and $\mathbf{v}_{\mathrm{p}}$ are the predicted position and velocity; $\mathbf{x}_0$, $\mathbf{v}_0$, $\mathbf{a}_0$ and $\dot{\mathbf{a}}_0$ are the position, velocity, acceleration and its time derivative at time t_0; and Δt is the timestep.

The corrector is given by the following formula (see, for example, [HMM95]):

$$\mathbf{x}_{\mathrm{c}} = \mathbf{x}_0 + \frac{\Delta t}{2}(\mathbf{v}_{\mathrm{c}} + \mathbf{v}_0) - \frac{\Delta t^2}{12}(\mathbf{a}_1 - \mathbf{a}_0), \tag{4.22}$$

$$\mathbf{v}_{\mathrm{c}} = \mathbf{v}_0 + \frac{\Delta t}{2}(\mathbf{a}_1 + \mathbf{a}_0) - \frac{\Delta t^2}{12}(\dot{\mathbf{a}}_1 - \dot{\mathbf{a}}_0). \tag{4.23}$$

The predictor formulae use only the "instantaneous" quantities that are calculated directly from the position and velocity at the present time. Compared to the scheme which has to keep track of values at previous timesteps, the program becomes much simpler.

The merit of the Hermite scheme is, however, not just the simplicity of the formula. The local trunction error of the Hermite interpolation is several orders of magnitude smaller than that of Newton interpolation with the same order and stepsize. Therefore, the Hermite scheme allows a significantly longer timestep than that used for the Aarseth scheme. Of course, this advantage is partially offset by the additional cost needed to calculate the time derivative directly. Thus, the relative advantage depends upon the computer used.

In the case of GRAPE systems, the Hermite scheme has a clear advantage over the Aarseth scheme, since the amount of work on the host computer is reduced substantially. First, the timestep can be nearly a factor of two longer for the same accuracy. In addition, the number of operations per timestep is much smaller for the Hermite scheme.

The HARP-1 system is the proof-of-the-concept system to demonstrate the feasibility of the special-purpose computer for the Hermite integrator. The pipeline is designed using floating-point processor chips like the LSI logic L64133 and Weitek 3364. The interface with the host is a VME. The design of the HARP-1 system started in spring 1992, and the system was completed by spring 1993. Eiichiro Kokubo designed and developed the HARP-1 hardware as part of his Master's thesis work.

The design of HARP-1 is similar to that of GRAPE-2. However, we were able to use a clock speed of 12 MHz, resulting in a speed of about 200 Mflops.

4.8   GRAPE-4

In 1990, we applied for a large grant to develop a big machine for the simulation of globular clusters using the Aarseth scheme. That year our proposal was rejected. However, in 1992, after we completed the GRAPE-3 system, the proposal was approved and we were awarded a 185M JYE grant (the total over five years) as a Specially Promoted Grant in Grant-in-aid from the Ministry of Education. Naturally, the machine to be developed by this grant was called GRAPE-4.

4.8.1   Architecture

The goal of the GRAPE-4 project was to develop a machine with a peak speed exceeding 1 Tflops by early 1995. The main target of this machine was the simulation of collisional systems. Therefore, the relative accuracy of the force had to be much higher than that adopted in GRAPE-3. In addition, the system should offer a reasonable performance for the individual timestep algorithm, applied to the system with a fairly small number of particles ($N \leq 10^5$).

The peak speed of 1 Tflops was not very difficult to achieve. Even with the technology available by 1990, we were able to implement one force calculation pipeline with mixed single- and double-precision accuracy into a single chip. This chip could operate at a speed of 30 MHz without much difficulty in design. Thus, one chip would offer a speed of about 1 Gflops. Thus, in order to achieve 1 Tflops, we needed to construct a machine with 1000 pipeline chips. The mass-production cost of the chip would be around 50K JYE per chip. Roughly speaking, the total cost of a machine with, say, 1500 chips would be around 150 M JYE, if we include design costs, etc.

To achieve a reasonable performance on the target application of globular clusters, however, the architecture of the total system had to be carefully designed.

The architecture of GRAPE-3 was quite simple (see Figure 39). One board consists of multiple pipeline chips which share one memory unit. Multiple boards are connected to a single I/O bus of the host. This architecture, however, is impractical for a machine with 1500 chips. Unless we adopt some exotic packaging, the total number of processor boards would be 50–100. This number of boards cannot fit into a single backplane bus.

Of course, we can connect an arbitrary number of boards to the host computer using a hierarchical structure, as shown in Figure 40. In the simplest case, we can use the bus repeaters in order to increase the physical number of boards to be attached.

However, we could not use this simple structure for GRAPE-4, the reason being that the number of pipelines was too large; the machine had more than 1000 pipelines. In the simplest case, different pipelines would calculate the forces on different particles. Thus, we had to calculate the force on more than 1000 particles in parallel in order to achieve an acceptable performance. One might think 1000 particles is not a large number compared to the total number of particles, since the target size of the number of particles of GRAPE-4 is 10^5. The average number of particles that share the same time is, however, not very large, since we are using the individual timestep algorithm. In practice, we have to design the hardware so that it runs efficiently for around 100 particles to share the same time.

If the data transfer speed of the host computer is very fast, we can still use the same structure and yet reduce the number of particles on which the forces are calculated in parallel. Consider the simplest case in which we have

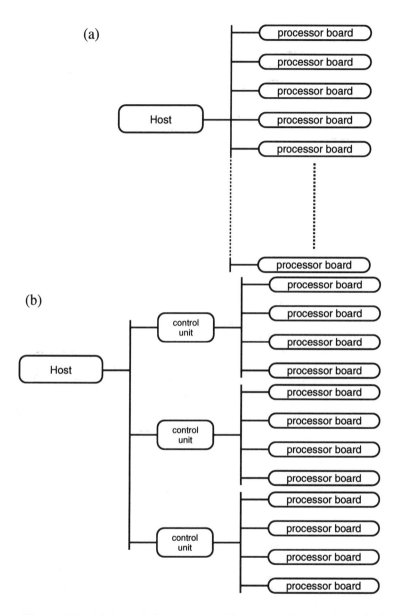

Figure 40   (a) A single-bus system. All processor boards share the host I/O bus. (b) A hierarchical system. Several boards form a cluster. Multiple clusters are connected to the host I/O bus.

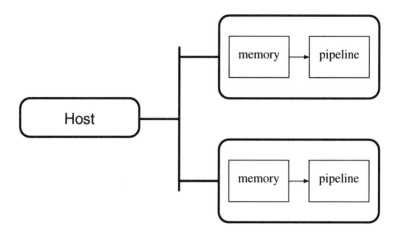

Figure 41   A simple multi-board system with two boards, each with one
pipeline unit.

two boards, each having one pipeline chip (Figure 41). The way we use the
hardware in the case of GRAPE-3 is to store the same data to both memories
of two boards, and let two boards calculate the forces on different particles.
The other possibility is to divide the particles into two subsets, send each
subset to different boards and let two boards calculate the force on the same
particle. After the calculations on both boards are complete, the host computer receives the results and sums the two results to obtain the total force
on a particle.

In the second approach, the number of particles for which the forces are
calculated in parallel is the number of pipelines on one board. However, the
amount of communication increases, since the host now receives two partial forces, instead of one. If we have many boards, the increase in the total
amount of communication becomes quite large. Roughly speaking, if we have
40 boards, the data transfer speed required would be around 500 MB/s, which
was possible only with expensive high-end vector processors.

In the case of the first approach, each processor board has its own copy of the
positions of particles stored in its memory. Thus, if the I/O bus supports only
point-to-point data transfer, writing the data to the memory would also take
a time proportional to the number of boards. In the case of GRAPE-3A, we
avoided this problem by letting the host broadcast the data. The idea of data
broadcasting is quite simple. Since all GRAPE boards are physically connected
to one bus, all boards can see the transaction (both data and address). Thus,
if the transaction is the data write to the memory, all GRAPE-3A boards
store the data in the VME bus to the memory, but only one board performs

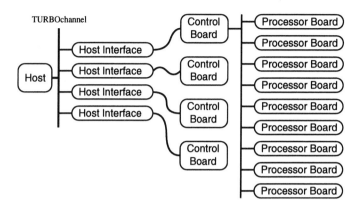

Figure 42 The top-level structure of GRAPE-4. (Reproduced from Makino *et al.* (1997).)

the actual handshaking. When we designed this broadcasting scheme, it was not defined in the VME specification. However, the newest revision of the VME specification includes this method of broadcasting. Thus, in the first approach the amount of communication does not depend upon the number of processors.

For the GRAPE-4 system, however, we had to use the second approach. Thus, we had to develop some way in which to reduce the cost of the communication to retrieve the calculated results from all boards. An obvious solution is to attach additional hardware which adds up the results from all the boards, so that the summation within a cluster is taken care of by this additional hardware.

Figure 42 shows the top-level structure of the GRAPE-4 system. The forces calculated on the processor boards under one control board are now summed by the control board itself, and the host has to sum only four partial forces, instead of 36 partial forces calculated on 36 processor boards. This feature reduces the requirement for the communication speed with the host to be around 100 MB/s.

The required transfer speed of 100 MB/s is more than 10 times higher than the speed that can be achieved by the PIO mode access of the VME bus. Moreover, even if we use the DMA mode on VME, the data transfer rate of more than 20 MB/s is practically impossible. Therefore, we chose to design an interface unit which plugs directly into some proprietary I/O bus of a workstation.

4.8.2 The host interface protocol

In the specification, many high-end workstations did offer a data transfer speed of around 100 MB/s. In early 1993, Sun SPARCstations, IBM RS/6000, SGI Indigo-2 and DEC Alpha AXP all offered speeds of around 100 MB/s. The performance close to the theoretical peak speed is, however, achieved only with DMA transfer, where the device accesses the main memory. Thus, for GRAPE-4, we had to give up the ease of programming using the PIO access, as well as the possibility of supporting multiple host computers through the use of a standard bus.

With the DMA I/O, the I/O bus can run at its peak theoretical speed if the I/O device is designed so that it can support the peak transfer rate. In the case of the interface to GRAPE, it is not very difficult to support the necessary bandwidth. To support the transfer rate of 100 MB/s, a parallel synchronous interface with a clock period of 25 MHz (40 ns) is sufficient. For the point-to-point transfer with a short wire length, 25 MHz is not a very high speed. In fact, the HIPPI standard sends a 32-bit parallel signal at a data rate of 25 MHz, for a link of up to 25 m.

However, to achieve this performance through an I/O bus is not very easy. The problem is the following: to achieve a reasonable data transfer speed, we had to use the DMA transaction, since the data transfer speed that can be achieved with PIO was inadequate. The use of DMA, however, implies that we have to do rather tricky programming to bypass the operating system overhead discussed in the section on GRAPE-2.

In normal use of the DMA operation, the user process first prepares the data buffer in its virtual address space. Then, it calls the kernel through a **read** or **write** system call. In the case of a read operation, the kernel then prepares its own data buffer in its address space and issues the command to the device to transfer data to that buffer area. Then, this process goes into a sleep state until the device sends a hardware interrupt to notify that the data transfer is complete. Then, the kernel process copies the transferred data into the user memory space and returns the control to the user process.

This procedure has three problems. First, the software overhead caused by the system call itself is pretty large. Second, the overhead associated with the interrupt handling routine is also significant. Third, the copy operation between the buffer in the user space and that in the kernel space requires considerable time.

To achieve high performance using DMA, we decided to develop software which bypasses the operating system completely. To bypass the operating system, the user process has to read/write the command registers of GRAPE-4 directly. Moreover, the DMA engine of the host interface has to access the physical memory allocated to the user process.

The direct access of GRAPE-4 command registers by the user process is implemented by the **mmap** function, as in the case of the VME interface of other

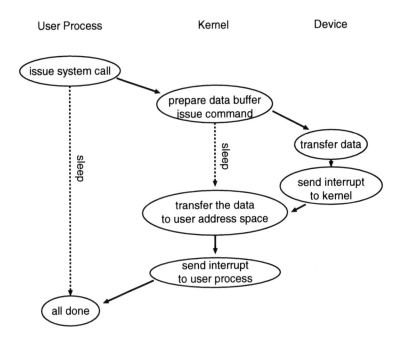

Figure 43 Normal DMA read operation on a UNIX operating system.

GRAPE systems. The only difference is that we have to develop the device driver software for the GRAPE-4 interface card. This is, however, relatively easy, provided the necessary documents are available from the workstation manufacturer (which is, in practice, not always the case).

Access of the user memory space by the DMA engine of the GRAPE-4 host interface is more difficult to implement, because the DMA engine needs to know the physical address corresponding to the virtual address within the user memory. The mapping between the physical address and the virtual address is done in the unit of the page size. Therefore, if the size of the data is larger than the page size, the physical address might be nonconsective, even if the memory address within the user virtual address is consecutive.

We chose the DEC Alpha AXP workstation with a TURBOchannel I/O bus as the primary host computer for the GRAPE-4 system. The problem with designing the system just for one proprietary interface is that the system cannot be connected to any other machine. Moreover, the computer manufacturer might change the bus to something else (which, unfortunately, had actually taken place by 1994, when DEC announced that their new machines would use the PCI bus).

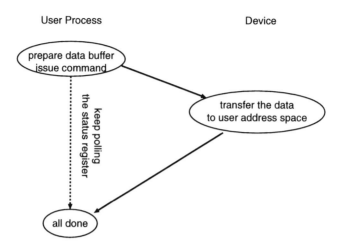

Figure 44   User mode DMA read operation used in GRAPE-4 interface
library.

We designed the total system so that only the host interface board depends
upon the particular selection of the host. In other words, GRAPE-4 can be
connected to the host computer with an I/O bus other than the TURBOchannel bus, if we design the host interface board for the I/O bus of that computer.
We have developed an interface board for PCI [KFTM97], which achieved a
communication performance comparable to that of the TURBOchannel implementation.

We chose the DEC Alpha AXP system for the following reasons. Firstly,
it offered quite decent floating-point and integer performance. The integer
performance is fairly important, since a significant fraction of the calculations
to be performed on the host computer are integer operations. Second, the
specification of the TURBOchannel bus was very simple, and therefore the
interface was relatively easy to develop. The TURBOchannel is a synchronous
I/O bus with a centralized arbitration mechanism. One unique feature of
the TURBOchannel bus is that each slot has dedicated control signal lines,
eliminating the necessity of the state machine to handle the bus arbitration.

In the usual implementation of the I/O bus, a DMA transfer would have
two phases. The first is the arbitration phase, in which the device requests
the right to use the bus. Since multiple devices may request the bus at the
same time, some mechanism to arbitrate between them is necessary. Only one
of the requesters actually gets the bus, and the others wait until the bus is
released. This means that the device should withdraw the request when it is
not granted, and retry when the bus is released again.

In the case of the TURBOchannel, the arbitration is not visible from the device. The device can simply request access and just wait until the host allows it.

The design of PIO operation is also simple. In a normal I/O bus, each device has its own address window. Thus, the device should receive the address and decode it to decide if it should respond. In the case of the TURBOchannel, the decoding is also performed by the host, and the request to the device is done by a single dedicated line.

These features of TURBOchannel, along with other features (or the non-existence of many features), made the design of the device particularly easy.

Finally, the Digital UNIX operating system, Digital's implementation of UNIX which runs on the Alpha AXP system, was reasonably stable and easy to modify.

4.8.3   The host interface board

Figure 45 shows the structure of the Host Interface Board (HIB). It consists of the data transceivers (trcv) to exchange data with the host and the control board, the FIFO (First-In First-Out) unit to buffer the data, and the control logics.

The host interface board should perform synchronous data transfer with both the host and Control Board (CB), despite the fact that host and CB operate on independent clock signals. To achieve this, the left-hand side of the board operates on the host clock and the right-hand side on the CB clock.

The size of FIFO unit is 2048 32-bit words (8kbytes). The data transfer rate between the host interface unit and the control unit is slower than that between the host and the interface board, because the clock frequency is lower. Using the FIFO buffer, the host computer can send the data at the peak transfer speed. In addition, there are several restrictions that the host interface board must obey for DMA transfer on the TURBOchannel bus. For example, the length of the single DMA burst cannot exceed 128 words, and the single DMA burst must not cross the 2KB address boundary. The FIFO buffer allows the control board to transfer data without testing the status of the host bus. As a result, the design of the control board becomes independent of the host bus.

The control logic is made of three 22V10 PLD chips and one AMD MACH230 complex PLD chip. The AMD MACH chip integrates most of the control logics, such as the DMA sequence controller, DMA word address counter, etc. Other PLD chips perform timing-critical handshake operations on both the TURBOchannel and the processor-board interface. We used Cypress CY7C453-14 clocked FIFO chips for the data FIFO. This clocked FIFO chip accepts separate clock signals for read and write data paths. The interface to TURBOchannel operates on the TURBOchannel clock, while the interface to the control board operates on the clock provided by the processor board. We used a Cypress CY7B991 PLL clock distributer with adjustable delay to distribute the clock signals.

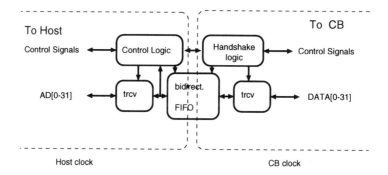

Figure 45 Host interface board. (Reproduced from Makino *et al.* (1997).)

4.8.4 The control board

The control board has two main functions: the first is to distribute the data received from the host computer to processor boards; the second is to retrieve the calculated force and potential from the processor boards and sum them up. The summed result is transferred to the host computer through the host interface board.

Consider the case in which we calculate the force from N particles to M particles. First, the host computer sends the position (and other) data of all N particles to the control board. The control board distributes the data to the processor boards, according to the indices of particles. Each processor board takes care of $N/9$ particles. Then, the host computer sends the position and velocities of M particles, in blocks of p particles, where p is the number of pipelines on a board. Each of the processor boards can calculate the forces on p different particles from the same set of particles. The control board broadcasts the data of these p particles to all processor boards. Therefore, all processor boards calculate the forces on the same p particles, from different set of particles. After all processor boards have finished the calculation, the control board sums up the partial forces calculated on the processor boards, and sends the result to the host computer.

The internal structure of the cluster is not visible from the application program. The host computer sends the data of p particles, and receives the calculated results for these p particles. The number of processor boards in one cluster changes the speed of computation, but nothing else.

In order to distribute the computation to different clusters, we use the same algorithm as that used for different processor boards in one cluster. If we have four clusters, the host sends $N/4$ particles to each cluster, where N is the total

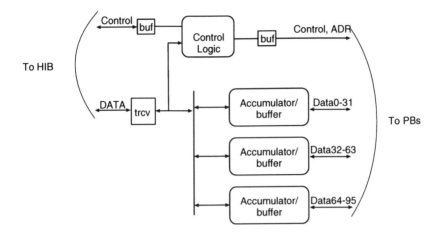

Figure 46   Control board. (Reproduced from Makino *et al.* (1997).)

number of particles that exert the force. In this case, each processor board takes care of the contribution of $N/36$ particles. Then, each cluster calculates the forces on the same p particles. The calculated result is summed up on the host computer.

The control board and processor boards are connected by a backplane bus (the HARP bus) with a 96-bit data width. A synchronous, pipelined protocol with a fixed latency is used on this HARP bus, where the control board always initiates the transaction. Therefore, there is no need for arbitration.

Figure 46 shows the structure of the control board. It consists of the control logic unit, three accumulator/buffer units, and several transceivers and buffers. The control unit generates all the necessary signals to control the HARP bus. The 96-bit data bus is divided into three 32-bit subbuses, each of which is connected to a different accumulator/buffer unit. The accumulator/buffer unit contains a 64-bit floating-point ALU, which accumulates the result calculated on the processor boards.

The control logic unit consists of three AMD MACH230 complex PLDs. The accumulator/buffer unit consists of a TI 74ACT8847 64bit floating-point LSI chip and transceivers, buffers and FIFO chips.

4.8.5   Processor board

Figure 47 shows the structure of a processor board. The particle data memory stores the data of the particles which exert the force. The PROMETHEUS LSI is used to calculate the position (and velocity) of particles at a specified time. This chip is a single-chip implementation of the predictor pipeline designed

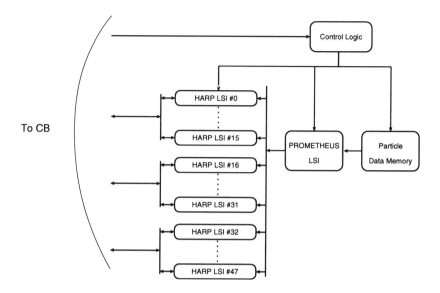

Figure 47   Processor board of GRAPE-4. (Reproduced from Makino *et al.*
(1997).)

for GRAPE-2. For GRAPE-4, this predictor pipeline is necessary to achieve
an acceptable performance. The HARP LSI chips calculate the gravitational
accelerations and their first time derivatives for particles. One board has 48
HARP chips.

The processor board also implements the neighbor list unit, similar to that
implemented in GRAPE-3.

4.8.6   HARP chip

Figure 48 shows the architecture of the HARP chip. It is simply a fully-
pipelined hardware implementation of formulae (4.16) and (4.17). We adopted
an architecture in which x, y and z components of all vector quantities are
processed sequentially, in order to reduce the gate count. Thus, it takes three
clock periods to calculate one interaction.

Each chip calculates the forces on two particles, using the "virtual multiple
pipeline" (VMP) [MKT93]. With VMP, the clock period of the pipeline LSI
is twice that of the system clock, and it calculates the forces on two different
particles at alternate clock cycles. From the outside, one force calculation
chip looks as if it has two pipelines. The advantage of this architecture is that
we can increase the performance of the pipeline chip without increasing the
system clock cycle.

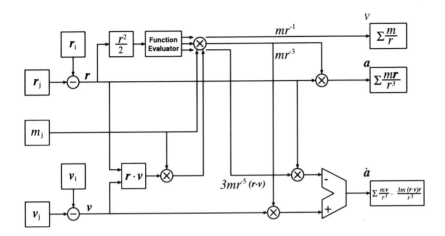

Figure 48 Block diagram of the HARP chip. (Reproduced from Makino *et al.* (1997).)

This approach is conceptually similar to what is used in "superpipeline" chips such as thte MIPS R4000 and "clock-doubled" chips such as the Intel i486DX2. In our VMP architecture, however, the chip actually has two separate sets of registers, and two virtual pipelines operate independently. On the other hand, in the case of chips such as i486DX2, the CPU still looks like one CPU. Only the cycle time of the external memory bus is reduced to a fraction of the internal clock cycle. The clock-doubling in the i486DX2 causes some performance penalty, since the relative speed of the main memory becomes slower. The penalty of the cache miss-hit increases significantly. Thus, it is not practical to further increase the ratio between the internal and external clock, unless the width of the memory bus is also increased.

It is also possible to compare our VMP with multi-threaded architectures such as the Denelcor HEP (Heterogeneous Element Processor, [Kow85]) or Tera Computer MTA (multi-threaded architecture). The processors of these machines have multiple register sets so that a fast processor looks like a collection of slow processors. Thus, the idea is similar to that of VMP. With a multi-threaded architecture, however, the data transfer rate of the main memory must still be fast enough to supply the data to a large number of "slow" processors. It is still not very easy to design the memory system. The multi-thread approach relaxes the requirememt for the latency, but does not change the requirement for bandwidth.

In the case of VMP, neither of the above problems limit the performance, since all virtual pipelines share the same input data. As a result, it is not very difficult to increase the number of virtual pipelines. In our architecture,

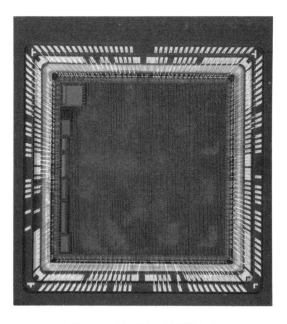

Figure 49   The HARP chip.

it is also easy to have physical multiple pipelines in one chip, when a larger number of gates becomes available. In the present HARP chip, there are two VMPs because the external clock speed of 16 MHz is already sufficiently low to make the design of the processor board easy. In addition, a further decrease in the clock cycle of the processor board would cause a decrease in the data transfer rate between the control board and the processor boards.

The subtraction of the position vectors and accumulation of the accelerations calculated are performed in 64-bit floating-point format. Other calculations to obtain the acceleration are performed in 32-bit format. The subtraction of velocity vectors and the accumulation of the time derivatives are performed in 38-bit format. Other calculations to obtain the time derivative are performed in 29-bit format, in which the length of the mantissa is 20 bits.

The HARP chip is cell-based. It is fabricated by LSI Logic Corp. The design rule is 1 μm and the gate count is about 100k. The die size is 14.2mm×14.2mm. The worst-case clock cycle of the chip is 30 MHz. We received the engineering sample of the HARP chip in August 1993. Figure 49 shows the actual die of the HARP chip.

Eight HARP chips are packaged in an MCM (Multi-Chip Module) package manufactured by Kyocera. As shown in Figure 47, 16 HARP chips share the same data bus. Therefore, all eight HARP chips in an MCM share the same input data bus and I/O bus. This design is well-suited for an MCM package because several difficult problems with MCM packages do not arise.

The use of MCM makes it possible to integrate a large number of HARP chips on a single processor board. As stated earlier, one processor board houses 48 HARP chips (six MCMs). If we use the normal single-chip package, it would be very difficult to integrate more than 16–24 chips on a single board. The total number of processor boards would be around 100 and the manufacturing cost wound be about 40% higher.

In general, MCMs are difficult to use because of the problems of testability and yield. Each LSI must be tested before actual use. However, the test for an MCM is difficult to design, because some of the signals cannot be set/sensed from outside the module. In the case of the HARP MCM, there is no such difficulty, since all I/O pins of HARP chips can be directly accessed from outside. Therefore, the test procedure for an MCM is the same as that for a single chip.

The yield of an MCM tends to be low, since the probability that an MCM works is the product of the probabilities that each chip in the module works. For example, if the yield of a single chip is 90%, the yield of an MCM with eight chips would be $(0.9)^8 = 43\%$. We expected the yield of the single chip after the DC test to be around 98%. In this case, the yield of the MCM is still higher than 80%. In other words, the yield is not a severe problem. In addition, we designed the processor board so that it can accommodate MCMs with defective chips, by adding the translation table between the logical chip number and physical chip number. Thus, if the processor boards have (at a maximum) two defective chips, we can use all the other 46 chips even if the physical locations of the defects are different on different boards.

The actual yield of MCMs with no defects was around 70%. This is slightly lower than our expectation, but good enough to allow us to assemble the required number of working boards. Figure 50 shows the bottom view of the MCM package with eight HARP chips mounted.

4.8.7 PROMETHEUS chip

The architecture of the PROMETHEUS chip is shown in Figure 51. It is a straightforward hardwired implementation of the predictor formulae (4.20) and (4.21). The subtraction of time and the addition of $\mathbf{x}_j$ and the higher order term are performed in 64-bit format. All other calculations are performed in 32-bit format. The PROMETHEUS chip handles x, y and z components sequentially, in the same way as the HARP chip.

The PROMETHEUS chip is a 1 μm CMOS gate array. It is fabricated by LSI Logic Corp. The raw gate count is about 182k. The actual gates used is about 60k. It is packaged into a 391-pin CPGA package.

Figure 50   Bottom view of the MCM package.

4.8.8   Development history

The development of GRAPE-4 started in summer 1992, when we were awarded the grant from the MoE. In fall 1992 we started the design of the HARP chip, and in winter 1992 the PROMETHEUS chip. These were completed in summer 1993, and a test board with a VME interface was developed. This board was completed in summer 1993, and we confirmed that both chips worked perfectly. A printed-circuit board version was developed based on this design and named HARP-2. Copies of this board are currently in operation at Cambridge University, Kiel University, Yamagata University and the Tokyo Institute of Technology.

In early 1993, we completed the conceptual design of the overall system, including the selection of the host computer. The prototype of the host interface board and control board were completed in January 1994 and the processor board in May 1994. The prototype system became operational in July 1994. The first batch of printed-circuit boards arrived in October 1994, and, after some debugging, a system of one HIB, one CB, and one PB with eight HARP chips becomes operational. The complete 36-board system became operational in July 1995. The largest fraction of the nine months between October 1994 and July 1995 was spent in waiting for delivery of the MCM packages from the manufacturer. It turned out that they had been a little too optimistic in estimating the production speed for the package.

Figure 52 shows a photograph of the completed GRAPE-4 four-cluster system with the host computer and one of the authors (J.M.). As can be seen from the photograph, GRAPE-4 is quite compact for a machine with a peak speed of 1 Tflops.

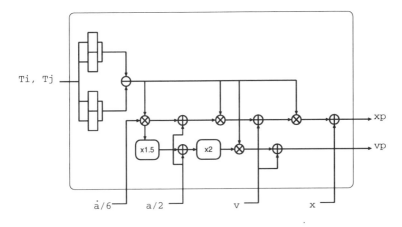

Figure 51   Block diagram of the PROMETHEUS chip. (Reproduced from Makino *et al.* (1997).)

Figure 52   The GRAPE-4 system with one of the authors (J.M.).

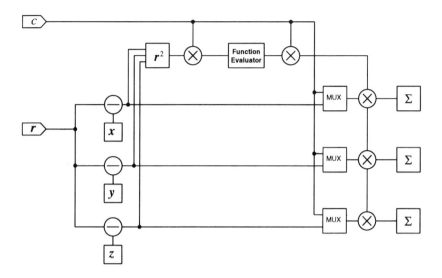

Figure 53   Block diagram of the MD-GRAPE chip.

4.9   MD-GRAPE

MD-GRAPE [FTM⁺96] is the successor of GRAPE-2A, which implements the pipeline unit of GRAPE-2A in a single custom VLSI chip. In the following, we briefly describe the MD-GRAPE system.

4.9.1   MD-GRAPE chip

The custom chip is fabricated by LSI Logic Corp. The design rule is 0.6 μm (cell-based ASIC, LCB300KL) and the gate count is about 400k. The worst-case clock cycle of the chip is 27 MHz. We received the engineering sample of the MD-GRAPE chip in early 1995. It can evaluate one interaction per clock cycle.

Figure 53 shows a block diagram of the MD-GRAPE chip. The pipeline of the MD-GRAPE chip evaluates one of the following equations:

$$
\begin{aligned}
\mathbf{f}_i &= \sum_j a_j g(b_j r_s^2) \mathbf{r}_{ij}, \\
\phi_i &= \sum_j a_j g(b_j r_s^2), \\
\mathbf{f}_i &= \sum_j \mathbf{p}_j g(\mathbf{k}_i \cdot \mathbf{r}_{ij}),
\end{aligned}
\tag{4.24}
$$

where $\mathbf{r}_{ij}$ is the relative distance vector, $g(x)$ is an arbitrary function, r_s^2 is the relative distance with the softening parameter ϵ, and the symbols a_j, b_j, $\mathbf{p}_j$, $\mathbf{k}_i$ are constants to determine atomic species, charges, masses, and so on.

The first and second equations are used to evaluate forces and potential energies, respectively. The last equation is used for the Ewald method, discussed in Section 4.5. Thus, the MD-GRAPE chip also integrates the functionality of the WINE pipeline. These three equations are calculated using the same arithmetic units. All calculations are performed in 32-bit single precision except for the calculation of relative position vectors and the accumulation of forces. The position vector is expressed by a 40-bit fixed-point number, and the force is accumulated in an 80-bit fixed-point number.

The main difference between the MD-GRAPE pipeline chip and the GRAPE-2A pipeline is in the function evaluation unit. The chip is designed so that Coulomb forces with a Gaussian kernel and atomic forces in Lennard–Jones potential have single precision accuracy with an appropriate cutoff. To combine the pipeline into single chip, we adopted 4th-order piecewise polynomial interpolation instead of linear interpolation. This change is necessary to reduce the size of the lookup table so that it can be integrated on a chip. The lookup table of GRAPE-2A has a total size of 8 Mbits, while that of the MD-GRAPE chip is about 100 kbits (the width of 98-bit × the depth of 1024 for the coefficients of the segmented polynomials, 8-bit × 1024 for the exponent table, and 6-bit × 1024 for the exponent correction). In addition, the data width and size of arithmetic unit for terms of different orders are optimized to minimize the circuit size. This difference is similar to the difference between the GRAPE-1/1A and GRAPE-3 pipelines.

The MD-GRAPE chip adopts a six-way VMP, so that we can reduce the input data bus of the chip. The operating clock of the pipeline chip is twice that of the data bus.

4.9.2   MD-GRAPE board

The overall structure of the MD-GRAPE board is essentially the same as that of the GRAPE-3 board, except that it has additional features such as an indirect addressing unit and a coefficient table lookup unit, which are described in Section 4.4.

The indirect addressing unit of the MD-GRAPE board is quite different from that in GRAPE-2A. Figure 54 shows the indirect address unit of MD-GRAPE. Instead of a single counter and lookup table, it has two counters. The operation, expressed in a pseudo Fortran code, would look as follows:

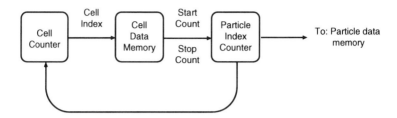

Figure 54   The indirect address unit of the MD-GRAPE board.

```
do isell=1, 27
    do iparticle=start_count(icell), end_count(icell)
        calculate force from particle  iparticle
    enddo
enddo
```

Particles in one cell are stored in consecutive locations of the particle data memory, and the host computer knows the first and last indices of particles for each cell. Before the start of the calculation, the host sends two arrays, start_count and end_count, with the first and last indices of the 27 cells. During the calculation, the cell data memory outputs the first and last indices of cell 1. The particle data counter counts from start_count(1) to end_count(1), and when finished, it sends the cell counter the request to count up.

To be precise, this signal is generated several clock cycles before the actual end of the first cell, so that the first index of the next cell is available to the particle data counter at the moment it reaches the end of the first cell.

With this double counter system, the amount of the data transfer is minimized. Moreover, the capacity of the cell data memory needed is much smaller than that of the index memory of GRAPE-2A.

The present MD-GRAPE board is a wire-wrapped board with a VME interface and Eurocard 6U form factor (card size is 233mm × 400mm).

4.9.3   Commercial activities

The MD-GRAPE chip was developed under a joint research contract between our group at the University of Tokyo and a commercial company called ITL (Image Technology Laboratory), with financial support from the government of Metropolitan Tokyo. ITL have developed two commercial versions of boards using MD-GRAPE chips, one with a VME interface and the other with a PCI.

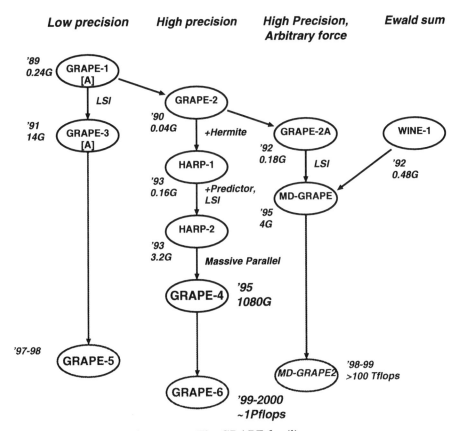

Figure 55   The GRAPE families.

4.10   Summary

Figure 55 shows the GRAPE hardware we have built so far and machines currently under development or in the planning stage. We have so far developed about 10 different machines in six years, and most of the machines are not just a techincal success but are intensively used for scientific research, both by our group and by researchers in other institutes. We believe this is a quite exceptional success for a highly specialized architecture such as GRAPE. In the next two chapters, we describe the software and algorithms used with GRAPE systems, and present some scientific results obtained so far.

We describe the machines under development in more detail in Chapter 7.

5

Software

In this chapter we describe the software for GRAPE systems. First we discuss the algorithms for the application program itself. Though many of the algorithms themselves can be found in the literature, the efficient implementation of them on GRAPE system requires somewhat different consideration from that for general-purpose computers. In addition, recent advances in the vectorization and/or parallelization, which are useful both on GRAPE systems and on general-purpose computers, are not well covered in the literature. Therefore, we decided to present a fairly detailed description of the algorithms and their implementation here.

First, we discuss the direct N-body methods, which are used mainly for collisional systems such as star clusters and planetesimals. Then we discuss the Barnes–Hut tree algorithm and other fast methods.

5.1 Direct methods

Here, we refer by direct methods to algorithms in which the force on a particle is calculated by actually taking a direct summation of the forces from all other particles in the system. The reader might well wonder what there is left to be discussed. If one uses the direct summation for force calculation, the algorithm to calculate force is given. Any time integration scheme can be used with the direct summation.

If very high accuracy is not necessary, the standard second-order leapfrog scheme is fine. If very high accuracy is necessary, either extrapolation methods or symplectic methods can be used. In fact, most textbooks on parallel scientific computing have this kind of attitude (see, for instance, [FWM94]). They describe the direct force calculation, and show that it can be easily parallelized for a pretty large number of processors. Some of the advanced books then describe more sophisticated (and approximate) force calculation algorithms such as the Barnes–Hut tree algorithm or FMM (Fast Multipole Methods), which will be discussed later in this chapter.

Unfortunately, real astrophysical N-body simulations are not that simple. In the following, we demonstrate the problem with the above simple-minded notion, and describe how astrophysicists tackled the problem.

5.1.1 An example problem

As an example, we consider a 100-body system. Particles are distributed so that they follow the mass distribution of the spherically symmetric, isotropic Plummer model, using the algorithm described in Aarseth et al. [AHW74]. Figure 56 shows a projection of the distribution of particles in the x-y plane. The system of units here is the standard unit [HM86a], where the total mass of the system M, the gravitational constant G, and the virial radius R_{vir} are all unity. Here, the virial radius is defined as

$$R_{vir} = \left\langle \frac{1}{|\mathbf{r}_i - \mathbf{r}_j|} \right\rangle^{-1}, \tag{5.1}$$

where $\mathbf{r}_i$ and $\mathbf{r}_j$ are the positions of particles i and j, respectively. Averaging is over all particle pairs $i \neq j$. In this system of units, the total energy of the system is $-1/4$ and the standard crossing time is $t_c = 2\sqrt{2}$. The virial radius is the harmonic mean of the distance of all particle pairs in the system, which gives a measure of the size of the system. Standard crossing time gives a typical orbital timescale of particles in the system. Note that if we fix M and R_{vir}, t_c does not depend upon the number of particles in the system, though the mass of individual particles, $m = M/N$, is smaller for a larger number of particles.

The mass density distribution of the Plummer model is expressed as

$$\rho = \frac{3M}{4\pi r_0^3} \left[1 + \left(\frac{r}{r_0} \right)^2 \right]^{-5/2}, \tag{5.2}$$

while r is the distance from the center of the cluster and r_0 is the scaling factor. In the standard unit, $r_0 = 3\pi/16 \simeq 0.589$.

Figure 59 shows the relative change of the total energy of the system integrated for three time units. Here, we used a six-stage, fifth-order RKF (Runge–Kutta–Fehlberg) scheme with a constant stepsize of $\Delta t = 1/512$. The energy jumps at several places.

These jumps are caused by close encounters of particles. In an N-body system, particles are attracting each other through gravitational interaction. Thus, if two particles happen to come close, they tend to come closer. Figure 57 shows gravitational close encounters of two particles with various velocities at infinity. All calculations start from coordinate $(-10, 1)$, with velocity $(v_0, 0)$. Depending on the initial relative velocity and impact parameter, the minimum distance between particles during an encounter can be arbitrarily small.

This characteristic is quite different from the behavior of atoms, where particles interact either with Coulomb force (repulsive) or van der Waals force (with potential minimum at a finite distance). Figure 58 shows the encounter of two particles with positive charges. The potential inhibits two particles

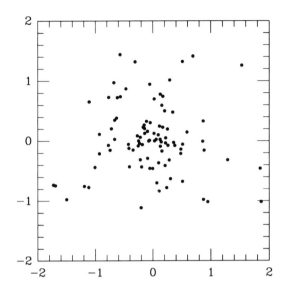

Figure 56   A 100-body *N*-body system with Plummer-model particle
distribution.

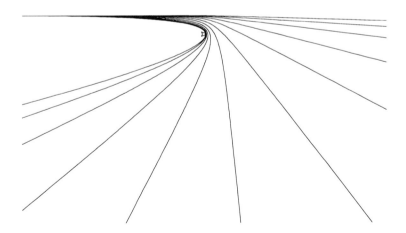

Figure 57   The trajectories of one particle in gravitational two-body
encounters.

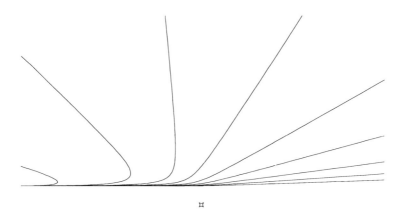

Figure 58   The trajectories of one particle in two-body encounters with
repulsive (Coulomb) potential.

from becoming too close. In both cases, the encounters with a low incoming
velocity result in a larger deflection. Thus, the physical effect is similar in
both cases. However, from the computational point of view, close encounters
under attracting interaction are much harder to handle.

An integration scheme with automatic stepsize control, such as the RKF
scheme, can reduce the error associated with a close encounter drastically. Figure 60 shows the energy change with automatic stepsize control. The energy
is now quite well conserved.

However, if we further continue the time integration of the system, we get
into trouble. Figure 61 shows the cumulative number of timesteps for the
time integration over 40 time units. In the case of the RKF scheme, the
slope becomes steeper as the calculation proceeds. In other words, the average
timestep shrinks. This is caused by the evolution of the system, which leads
to an increase in the central density (see Section 6.2). As the cluster evolves,
the central density and the central velocity dispersion become higher. As a
result, the timestep must shrink in order to maintain accuracy.

This increase of the central density continues until the energy production by
binaries (see Section 6.2) halts the contraction. The numerical techniques to
handle this phase will be discussed later. Here we describe the basic method to
handle the evolution of the system until the formation of binaries. As will be
seen in Chapter 6, this basic scheme is very powerful to handle a wide variety
of gravitational many-body systems. The dashed curve shows the number of
timesteps for this scheme, which will be discussed in more detail in the next
section.

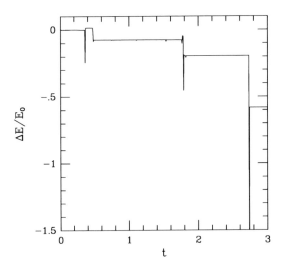

Figure 59 Time variation of the total energy of the system. Integration method is a six-stage, fifth-order Runge–Kutta–Fehlberg scheme with constant stepsize.

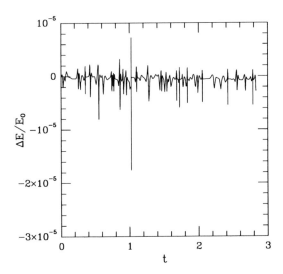

Figure 60 As Figure 59 but with automatic timestep adjustment.

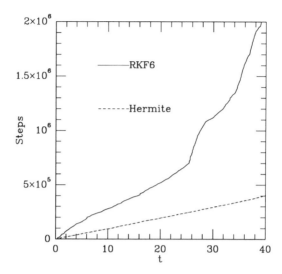

Figure 61 Number of timesteps as a function of time. Solid curve indicates
the result for the RKF scheme with shared timestep, and dashed curve the
Hermite scheme with individual timestep.

5.1.2 The individual timestep algorithm

The individual timestep scheme [Aar63], [Aar85] has been the only algorithm
that can be used for simulations of gravitational many-body systems, such as
open clusters, globular clusters and a system of planetesimals. In a simulation
of these systems, we are interested in the change of orbit of each particle due
to gravitational encounters with other particles, and the evolution of the total
system driven by such changes. In that sense, simulation of these systems is
similar to the molecular dynamics simulation, in which we are interested in
the thermodynamical process.

In simulations of these collisional systems, we need to follow the change
of orbit due to individual encounters with a reasonable accuracy, since the
encounters drive the evolution of the system. Therefore, we cannot use the
softening parameter to simulate the evolutions of these systems. When we use
a strict $1/r$ potential, the force between two particles changes very rapidly
when two particles undergo a close encounter. Therefore, during a close encounter, the timestep should be sufficiently small to resolve this rapid change.
Roughly speaking, the timestep is determined by the distance to the nearest
neighbor. Therefore, if we integrate the system in lockstep, the timestep is determined by the pair of particles with minimum separation. Even in a nearly
homogeneous system, the minimum separation is proportional to $N^{-2/3}$, and

the dependence on N is stronger when the system is highly inhomogeneous [MH88].

To reduce the total calculation cost, Aarseth [Aar63] developed an algorithm which assigns each particle its own timestep. In this individual timestep algorithm, each particle adjusts its timestep so that it satisfies the required accuracy. Thus, when two particles undergo a close encounter, although the timesteps of these particles shrink as required, the timesteps of other particles remain long. Makino and Hut [MH88] showed that the individual timestep scheme is faster than a shared timestep scheme by a factor in the range of $O(N^{1/3})$ and $O(N)$, depending on the distribution of particles.

In the individual timestep algorithm, each particle has its own timestep, and therefore its own time. To calculate the force on a particle due to other particles, we must know their positions at the time of the particle for which we calculate the force. To calculate the position of a particle at a time different from the time of the particle, Aarseth uses a third order polynomial extrapolation. This polynomial is evaluated each time a pairwise force is calculated, therefore the calculation cost of the polynomial evaluation is of the same order as that of the force calculation itself. The basic algorithm looks like the following in the case of Aarseth's program, which uses a predictor-corrector scheme:

(a) Select particle i with a minimum $t_i + \Delta t_i$. Set the global time (t) to be this minimum, $t_i + \Delta t_i$.

(b) Predict the positions of all the particles at time t using the extrapolation polynomial.

(c) Calculate the acceleration $(\mathbf{a}_i)$ for particle i at time t, using the predicted positions.

(d) Apply the corrector for the position and velocity of particle i.

(e) Go back to step (a).

Figure 62 illustrates how the individual timestep scheme works. When particle i is integrated from t_i to $t_i + \Delta t_i$, the positions of all other particles are predicted at that time, and forces from those particles are calculated using these predicted positions.

Not all time integration schemes can be used with this individual timestep, since the predicted values of the positions of all other particles need to be calculated at the time of the particle to be integrated. Most integration schemes do not provide the solutions at arbitrary points in time. For example, there is no simple way to obtain the approximate solution at the intermediate time with the usual Runge–Kutta schemes or extrapolation schemes.

As described already in Sections 4.3 and 4.7, predictor-corrector multistep integrators with a variable stepsize are best suited to the individual timestep algorithm.

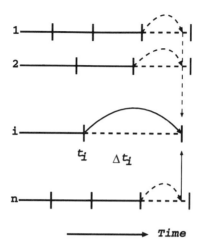

Figure 62 Schematic description of the individual timestep algorithm.

5.1.3 The hierarchical timestep algorithm

If several particles share exactly the same time, we must predict the position of the other particles only once in order to calculate the forces on these particles. The cost of the prediction therefore becomes a small fraction of that of the force calculation. In addition, the program becomes parallelizable more efficiently, since both the force calculation and the time integration can be done in parallel for particles with the same time.

In order to force several particles to share exactly the same time, we adjust the timesteps to integer powers of two. By this modification, all particles that have the same timestep also have the same time. We call this scheme the hierarchical timestep algorithm, since the timesteps of particles are hierarchically organized to powers of two. This algorithm was developed by McMillan [McM86] in order to use the Cyber 205 more efficiently. The technical details concerning this algorithm have been described by Makino [Mak91a].

On average, the number of particles which share exactly the same time is $O(N^{2/3})$, if the system is nearly homogeneous. The reason is explained as follows. The timestep of an average particle is $O(N^{-1/3})$, since the stepsize is inversely proportional to the average interparticle distance. On the other hand, the minimum timestep is $O(N^{-2/3})$ [MH88]. Therefore, the total number of timesteps per crossing time is $O(N^{4/3})$, and the number of timesteps of a particle with the minimum timestep is $O(N^{2/3})$. At each minimum timestep, therefore, $O(N^{4/3}/N^{2/3}) = O(N^{2/3})$ particles are updated, on average. As long as the system is nearly homogeneous, the hierarchical timestep scheme is very efficient. Half of the theoretical peak speed of GRAPE-2 is obtained for $N \simeq 500$ [Mak91a].

The implementation of this hierarchical timestep algorithm on GRAPE is quite simple. We are currently using Aarseth's NBODY1 and NBODY3. In order to implement the hierarchical timestep algorithm, we must modify only two subroutines in the case of NBODY1. The modifications needed to use GRAPE are also limited to one subroutine.

5.1.4 Second-order scheme

The individual timestep algorithm is useful for collisionless systems with a large density contrast. The cosmological N-body simulation of a highly non-linear stage is one such example. Another example is a realistic galaxy model. Even in the case of an elliptical galaxy with a relatively large core, the core mass is of the order of 1% of the total mass, which means that the core density is more than 100-times greater than the average density.

The individual timestep algorithm is a powerful tool for simulating such systems. In this section, we describe the implementation of a second-order individual timestep scheme that can be used with GRAPE hardware with low accuracy. This code was used to simulate the evolution of massive black-hole binaries in elliptical galaxies [MFOE93].

We use a second-order predictor-corrector scheme as the basic integrator. The predictor is given by

$$\mathbf{x}_p = \mathbf{x}_0 + \Delta t \mathbf{v}_0 + \frac{1}{2} \Delta t^2 \mathbf{a}_0, \qquad (5.3)$$

while the corrector is given by

$$\mathbf{x}_1 = \mathbf{x}_p, \qquad (5.4)$$

and

$$\mathbf{v}_1 = \mathbf{v}_0 + \frac{1}{2} \Delta t (\mathbf{a}_0 + \mathbf{a}_1). \qquad (5.5)$$

Here, subscript 0 denotes the value at time t, and subscript 1 denotes the value at time $t + \Delta t$. The corrector is a simple trapezoidal scheme with second-order accuracy. Note that this scheme is equivalent to a standard leapfrog, when a constant stepsize is used.

To use the GRAPE hardware efficiently, it is necessary to avoid the prediction of positions of other particles, since GRAPE hardware, except for GRAPE-4, does not have hardwired predictors. We can reduce the prediction cost significantly by using the positions of particles at $t_i + \Delta t_i/2$ to calculate the force at intermediate times [MF93]. Figure 63 shows how this prediction at halfway works in practice.

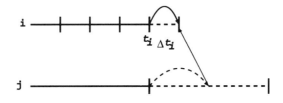

Figure 63   Schematic description of the second-order individual timestep
algorithm.

5.2   The treecode

Pfalzner and Gibbon [PG96] devoted a whole book to a description of the
treecode. Here we give a brief description and present the optimization techniques specific to GRAPE systems.

The basic idea of the treecode [BH86] is to replace the force from a group
of distant particles by the force from their center of mass or by a multipole
expansion. To ensure accuracy, we make groups for distant particles large and
groups for nearby particles small.

We use a tree structure to construct the appropriate grouping for each
particle. Before calculating the forces on particles, we first organize particles
into a tree structure. Barnes and Hut [BH86] used an oct-tree based on the
recursive subdivision of a cube into eight subcubes. We stop the recursive
subdivision if the cube has only one, or no, particle. See Makino [Mak90] for
details concerning an efficient tree construction algorithm. Figure 64 shows
the Barnes–Hut tree in two dimensional space.

After the tree is constructed, for each node of the tree, which corresponds to
a cube of a certain size, we calculate the coefficient of the multipole expansion
of the gravitational force exerted by particles in that cube. A fast algorithm
was described by Hernquist [Her90]. In the case of the treecode with GRAPE,
we use only the monopole term, since GRAPE can calculate only the force
from a point mass.

The force calculation is expressed as a recursive procedure. To calculate the
force on a particle, we start from the root node, which corresponds to the
total system. We calculate the distance between the node and the particle (d)

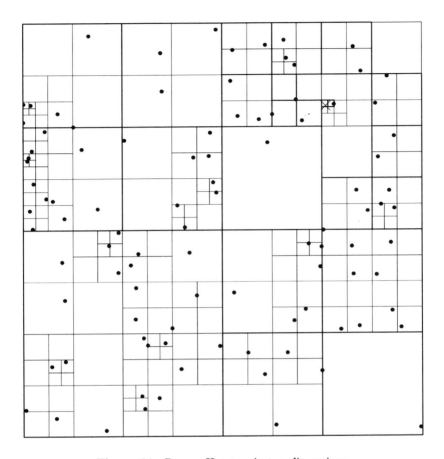

Figure 64   Barnes–Hut tree in two dimensions.

and compare it with the size of the node (l). If they satisfy the convergence criterion

$$\frac{l}{d} < \theta, \tag{5.6}$$

where θ is the accuracy parameter, we calculate the force from that node to the particle using the coefficients of the multipole expansion. If criterion (5.6) is not satisfied, the force is calculated as a summation of the forces from eight sub-nodes.

Usually, we use the distance between the particle and the center of mass of the node to determine whether the force is accurate enough. When θ is very large, this criterion can cause unacceptably large error [SW94]. For most calculations, however, such a pathological situation is not realized. In addition,

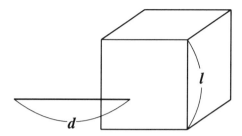

Figure 65   Opening criterion for tree traversal.

for treecode with GRAPE, relatively small values of θ such as 0.5–0.6 are not too costly.

It is difficult to accelerate the above-mentioned original implementation of the tree algorithm by GRAPE, since the calculation cost of the force calculation and tree traversing is of the same order. To determine whether to subdivide a cube or not, we must calculate the distance between the particle and the cube. This calculation must be performed on the host computer, since we need a distance of just one node. We must therefore calculate the distances to all nodes on the host computer.

We have developed an algorithm to use the treecode on GRAPE with high efficiency [Mak91c]. The basic idea is to construct a list of nodes that can be used to calculate the forces on many particles. If we can construct such a list, we can accelerate the force calculation using GRAPE. This algorithm was first developed by Barnes [Bar90] in order to vectorize the force calculation on a CDC Cyber 205. The list of nodes is constructed for a group of particles which are close to each other. The tree structure is used to construct groups of an appropriate size.

Figure 66 illustrates the modified opening criterion for this algorithm. Instead of calculating the distance between one box and one particle, we should calculate the minimum of the distances of all particles in a box to the other box. We use the geometrical minimum distance between the box and the center of mass of the other box as the distance [Bar90].

The overall computing speed depends upon the choice of the group size. If we use large groups, the calculation cost on the host computer decreases, since the number of tree traversals decreases. On the other hand, the calculation cost on GRAPE increases, because the length of the list becomes longer as we increase the size of the group. The list becomes longer for the following reasons: (a) the list contains particles in the group itself; (b) the nodes close to the group are subdivided to finer levels as we increase the size, since the node must satisfy criterion (2) for all particles in the group. Therefore, there

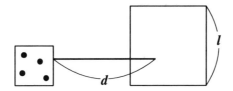

Figure 66   Modified opening criterion for treecode with GRAPE.

is one optimal group size, depending on the ratio between the speed of the host and that of GRAPE. The choice of the group size and the performance obtained has been discussed in detail by Makino [Mak91c].

5.3   Interface software

In this section we summarize how GRAPE is controlled by the host computer from the point of view of low-level hardware/software. First we describe the interface for machines other than GRAPE-1 and 4. All GRAPE hardware other than GRAPE-1 and 4 are accessed through a VME bus, as a slave device. Then we describe how these machines are actually used in the application program. Finally, we discuss the interface design of GRAPE-4.

5.3.1   Memory-mapped interfaces

In the case of a VME-based GRAPE, all of the memories and registers of GRAPE hardware which the host needs to access are mapped to the VME address space. Figure 67 shows a typical configuration of VME-based GRAPE with the host computer.

Some of the relatively old UNIX workstations used VME bus as the I/O bus, but most recent workstations adopted either a proprietary I/O bus such as the Sun SBus or DEC TURBOchannel, or PCI, in order to reduce the physical size of the system and the power consumption. However, we can connect a board with a VME interface to almost any workstation, since there are several companies which manufacture special interface hardware between various I/O buses and the VME bus. VME bus is a widely used standard, therefore the market for these interfaces is relatively large and these products are not very expensive. Table 6 summarizes the VME bus interface products we have used so far.

These interfaces transform the I/O operation on the host I/O bus to that on the VME bus. Figure 68 schematically shows how the transformation is

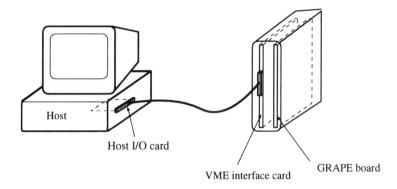

Figure 67   A VME-based GRAPE system.

Table 6   Bus adapter products for VME bus.

Manufacturer	Host bus	Hosts
Solflower	SBus	Sun SPARC
Bit-3	SBus	Sun SPARC
	PCI	Intel, DEC
Aval Data	SBus	Sun SPARC
National Instrument	MicroChannel	IBM RC/6000

performed. Typically, an I/O card on the host bus occupies a certain range of the physical address space of the host CPU. In some cases, each card slot has its own dedicated range of the address space, and in others, "plug-and-play" power-on setup software assigns the necessary address space. The host CPU can access this space by the usual load/store operations.

The bus adapter hardware maintains the mapping from the host address space to the VME space. In the example shown in Figure 68, both the host address space and VME address space are 32 bits. The access to the address 0xbfxxxxxx in the host physical address space is translated into the access to the address 0x00xxxxxx in the VME address space.

The UNIX system kernel takes care of the address mapping between the virtual address space of the user program and the physical address assigned to the bus adapter. On some machines, the address space assigned to an adapter slot is accessible through a predefined device special file. On some other machines, a simple device driver with support for memory mapping has to be developed.

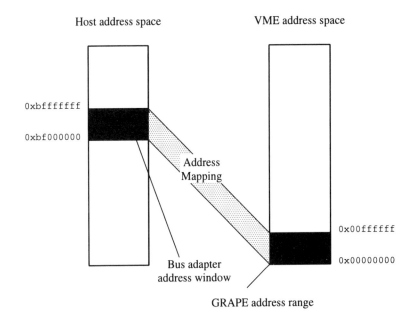

Figure 68 Address mapping with VME bus adapters.

Once all these have been done, the user process can directly access the physical memory and registers of GRAPE. For example, to store particle data to the memory or a register, the host CPU simply performs a store operation. To read the calculated force and other results, it performs a load operation.

Figure 69 shows the calculation flow in the case of the GRAPE-2 system. Here we consider the case where we calculate the force from all particles to all particles, using simple direct summation. The host computer first sends parameters such as the softening size, and then stores the data of all particles in the memory of GRAPE-2. Then the host computer stores the position of a particle in the registers of GRAPE-2, and sends the start signal. A dedicated address is assigned to this start signal, and GRAPE-2 starts calculations when this address is accessed.

After issuing the start signal, the host computer reads the status flag register. While performing calculation, GRAPE-2 returns "1" for the access to this register, and returns "0" when the calculation is finished. Thus, the host keeps reading the status flag register until it changes its value to "0".

This is the same polling (spinning) as we described in Section 4.2. The standard technique to delay for a peripheral device to finish its work is the hardware interrupt, which means the GRAPE-2 board drives a dedicated signal line (interrupt line) to notify the end of calculation to the host computer.

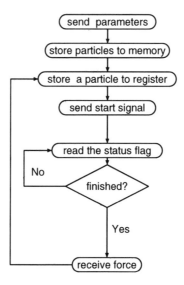

Figure 69 Calculation flow for the direct N^2 algorithm on GRAPE-2.

In this approach, the host computer can run some other processes while waiting for the end of calculation. On the other hand, in the case of polling, all of the CPU resource is wasted just to read the status flag.

We decided not to use the interrupt for the following reasons. The first reason is that the interrupt-based approach increases the software overhead, because the interrupt signal is taken care of by the operating system. When the interrupt is generated, the control is first transferred to the operating system kernel from the user process which was active at that moment. Then the kernel determines the source of the signal, changes the state of the process which uses GRAPE, and returns control to a user process. This rather complex procedure does require quite a long time, of the order of 1 millisecond on the 68030-based workstation we used as the host of GRAPE-2. Of course, on newer workstations this overhead is shorter, but the overall speed of both GRAPE and the host are improved by even larger factors. Thus, the relative cost of the overhead of the operating system is larger on modern machines.

The second reason for not using the interrupt is that the software would become more complex and less portable. As described above, the interrupt is taken care by the operating system. Therefore, we have to modify the operating system software to let it handle the interrupt signal properly.

We have used a number of workstations as the host computer. Some of them use the VME bus as the system's I/O bus (Sumitomo Electric Sumistation S-301 and Silicon Graphics Crimson). Some do not have a VME bus, but can be connected to the VME bus through "bus adapters" (Sony NEWS, Sun SPARCstations, IBM RS/6000, DEC Alphastations). In both cases, the user program can directly access the physical VME address space by mapping the VME address space into the virtual address space of the user program through an **mmap(2)** system call of the UNIX operating system.

The part of the code that manipulates GRAPE is machine-independent. The part of the code which sets up the bus-adapter is machine-dependent, since we must set up each adapter using the software supplied with that particular adapter. This machine-dependent part, however, can be encapsulated into a single subroutine, which can be used for any GRAPE hardware. Therefore, porting the software to a different host computer is a relatively simple task.

The fact that porting is easy does not mean that selection of the host computer is not important. When we choose a host computer, we must consider various issues. The speed of the floating-point operation, the speed of data transfer through the VME bus, and the reliability of both the operating system and the compiler are all very important in selecting a host computer.

In the case of a VME-based system, we feel the reliability of the host computer and the bus adapter is the most important factor, since any difference in the reliability can cause a vast difference in the productivity. If the host computer is a factor of two slower, the total performance of the system becomes degraded by a factor of two in the worst case. If the machine crashes every one or two days, the loss in computer and human time would be much greater than a factor of two.

For VME-based GRAPE systems, UNIX-based computers are not the only option for the host computer. For example, we can attach bus adapters to personal computers with Intel x86 CPUs or a Macintosh. It is also possible to use a VME-based one-board computer as a host. So far, we have not used these machines as the host computer, because their CPU performance was lower than that of workstations, and because the software for these systems was less reliable.

Recently, the difference in the performance of RISC-based workstations and Intel-based PCs or the PowerPC-based Macintosh has diminished. So these machines, in particular with a robust UNIX operating system such as Linux and FreeBSD, might offer a more cost-effective solution.

5.3.2 Application interface

In this section we describe the basic interface subroutines using examples. The main body of the Fortran code that calculates the force on all particles in an N-body system using GRAPE-2 [IEMS91] or GRAPE-2A [IMF$^+$93] looks as follows:

```fortran
real x(3,n), mass(n), acc(3,n), pot(n)
...........
call g2init
call g2setn(n)
do j = 1, n
    call g2setxj(j-1,x(1,j))
    call g2setm(j-1,mass(j))
enddo
do i = 1, n
    call g2frc(x(1,i),i,eps,acc(1,i),pot(i))
enddo
call g2free
```

This code actually works on both GRAPE-2 and GRAPE-2A, though the hardware is quite different. For comparison, the program to calculate the gravitational force on a general-purpose computer would look as follows:

```fortran
real x(3,n), mass(n), acc(3,n), pot(n)
do k =1,3
    do i=1,n
        acc(i,k)=0.0
    enddo
enddo
do i=1,n-1
    do j=i+1,n
        do 50 k = 1,3
            dx(k)=x(k,j)-x(k,i)
        enddo
        r2inv=1.0/(dx(1)**2+dx(2)**2+dx(3)**2+eps2)
        r3inv=r2inv*sqrt(r2inv)
        do k = 1,3
            acc(k,i)=acc(k,i)+r3inv*mass(j)*dx(k)
            acc(k,j)=acc(k,j)-r3inv*mass(i)*dx(k)
        enddo
    enddo
enddo
```

Note that the program which uses GRAPE-2 is actually shorter than that which runs on a general-purpose computer. In certain cases, we can even use a single subroutine call which takes care of everything.

```
call accel_by_grape2_separate(n, x, n,x, mass,acc, phi,eps2)
```

In the following, we briefly describe the functions of each procedure. We call particles which are stored in the memory of GRAPE hardware j-particles, and particles on which the forces are calculated i-particles.

The g2init procedure initializes the GRAPE-2 hardware. It's basic function is to map the VME address space assigned to GRAPE into the virtual memory space of the user program. In addition, it performs access control. If someone else is currently using GRAPE, it waits until GRAPE is released.

The g2setn procedure stores the number of j-particles (N), and g2setxj and g2setmj store the position and mass of a j-particle x_j and m_j, respectively. Though the user program can directly access the memories and registers of GRAPE, it usually accesses them only through these library procedures. This approach greatly simplifies the development of the user program and the task of porting the user program from one GRAPE system to another. Table 7 summarizes these interface library functions. In this book, all sample program fragments are shown in Fortran, but actual library functions are all written in the C language, and are used with application programs written in C or C++ as well.

In the case of GRAPE-2/2A, g2setxj and g2setmj simply copy the arguments to the specified address in the VME space, since they use the IEEE-754 number format, which is the same as that used in almost all workstations. However, those machines used for collisionless simulations, such as GRAPE-1A and GRAPE-3/3A, use a number format that is different from that of the IEEE-754 format. GRAPE-1A uses a 16-bit fixed-point format for the position, and GRAPE-3/3A use a 20-bit fixed-point format. However, the user program does not have to convert the data format, since the interface subroutines convert the data format before storing on these machines.

The g2frc procedure stores the position of the i-particle to GRAPE, issues a start signal, polls the flag register of GRAPE until the calculation is completed, and reads both the calculated force and potential. Here, again, the format conversion is taken care of within this subroutine in the case of GRAPE-1A and GRAPE-3/3A.

In the case of the $O(N^2)$ direct summation algorithm, we could use a subroutine that receives the positions and masses of all the particles and then calculates the forces on all of the particles that it received. We provide low-level procedures such as g2frc to allow a more flexible control of the GRAPE hardware.

Table 7   GRAPE-2/2A interface library subroutines.

Name	Args	Type	Description
g2init	—		acquire GRAPE hardware
g2free	—		release GRAPE hardware
g2setn	n	integer	number of particles
g2setxj	j	integer	location in memory
	x	real vector	position of a particle
g2setmj	j	integer	location in memory
	m	real	mass of a particle
g2frc	x	real vector	position of a particle
	i	integer	index of particle to skip calculation
	eps	real	softening parameter
	acc	real vector	calculated force
	pot	real	calculated potential

5.3.3   Message-based interfaces

The interface for GRAPE-4 is radically different from that of VME-based GRAPE systems, because we used the DMA transfer to achieve high performance.

Table 8 summarizes the data transfer operations between the host and GRAPE-4. To the host computer, GRAPE-4 looks like a peripheral device such as a hard disk unit. The host issues commands to GRAPE-4 and reads its status using PIO access, and GRAPE-4 (to be precise, the host interface board) transfers the data using DMA. For example, the procedure to store particles to the memory is expressed as follows:

(a) Store the packet data for particles in the DMA buffer area of the main memory. One packet consists of 19 words.

(b) Write the word count and physical start address of the DMA buffer to the DMA registers of the host interface board. This is done using a PIO write.

(c) Send the command to the control board. This command contains the number of particles as the parameter.

(d) The control board requests the host interface board to initiate a DMA read operation and send the received data.

(e) The host interface board reads the DMA buffer and sends the data through FIFO.

(f) The control board receives the data and stores the particle packets to the memory.

Table 8   Data transfer operations of GRAPE-4.

Function	Operation mode	Word count
Command issue	PIO write	2 (address + data)
Memory write	DMA read	up to 2048
Chip data register write	DMA read	960
Chip data register read	DMA write	960
Status read	PIO write	960
Neighbor list read	DMA write	up to 2048

The calculation flow on GRAPE-4, in the case of the hierarchical timestep scheme, is shown in Figure 70.

The application interface of the GRAPE-4 interface library is largely similar to that of GRAPE-2/2A, except that subroutines are designed to send/receive multiple particles.

5.4   Timesharing

Typically, GRAPE hardware is used by several people. It is therefore strongly desirable that GRAPE hardware can run multiple programs simultaneously. Even in the case in which only one person is using the GRAPE hardware, she or he might want to run several programs simultaneously. Consider the case in which we are running a large production run which would take several hours or even days. It would be very convenient if we could run other programs without stopping the production runs, since we could develop and test other programs while the production runs continue.

We adopted a very simple approach to achieve such timesharing. The user program acquires the GRAPE hardware when it calls the gXinit subroutine (X represents the machine types, such as 1a, 2, 2a, 3 and 3a) and releases it when it calls the gXfree subroutine. Figure 71 shows how multiple programs share one GRAPE hardware. First, program A acquires GRAPE by calling gXinit and starts calculations. After that, program B calls gXinit. The subroutine gXinit checks whether someone else is already using GRAPE or not. If so, it waits until the GRAPE hardware is released. Program A releases GRAPE by calling gXfree after using it for a certain period. Then, program B acquires GRAPE and starts to use it.

For this "timesharing" scheme to work, each user program must release GRAPE sufficiently often. At our site, it is generally recommended to release GRAPE hardware once every one or two minutes. This period was determined as a compromise between the machine's efficiency and the comfort of the programmer. If the period is too short, the efficiency of the machine decreases,

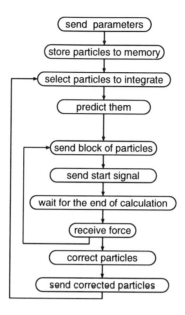

Figure 70   Calculation flow for the hierarchical timestep method on
GRAPE-4.

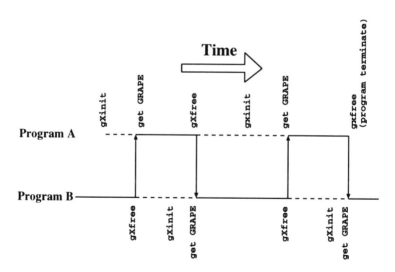

Figure 71   Timesharing between two programs using the same GRAPE
hardware.

since the overhead of init/free routines becomes larger. If the period is too long, it becomes painful for a user to wait for the test program to run. It is straightforward to add the necessary code to release/acquire GRAPE after a certain period.

Note that this simple approach works fine because the amount of data which resides on GRAPE hardware is small, and the communication between the host computer and GRAPE is relatively fast. On many dedicated computers, it turned out to be difficult or impossible to run multiple programs simply because the checking in and out of the data to some archive would take too much time.

5.5   Optimization

In this section, we discuss several optimization techniques which give some performance improvement. Problem-specific techniques will be discussed in more detail in Chapter 6. Here we concentrate on more general aspects of taking advantage of the parallelism.

5.5.1   Concurrent operation in multi-board systems

Both GRAPE-3 and GRAPE-4 consist of multiple boards or multiple clusters which are connected to a single I/O bus of the host computer (see Figure 40). In these systems, a naïve implementation of the force calculation with $O(N^2)$ direct summation would look as follows:

```
real x(3,n),m(n),a(n)
...............
do j=1,n
   call g3setxj(x(1,j),j)
   call g3setmj(m(j),j)
enddo

do i=1,n,nchips
   np=min(nchips,n-i+1)
   do ichip=,nchips
      call g3setxi(x(1,i+ichip-1), ichip)
   enddo
   do iboard=1,nboard
      call start_board(iboard)
   enddo
   do iboard=1,nboard
      call wait_board(iboard)
```

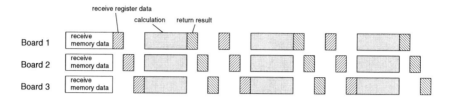

Figure 72 Parallel operation on a multi-board system with a naïve
implementation.

```
   enddo
   do ichip=0,nchips-1
      call get_force(a(1,i+ichip), pot(i+ichip),ichip+1)
   enddo
enddo
```

Here, x, m and a are the position, mass and calculated gravitational acceleration, respectively. The first loop stores the particle data to the memory of
each board. Subroutines g3setxj and g3setmj store data to all boards. Subroutine g3setxi writes the position of a particle on the on-chip register of the
force calculation pipeline chip. Here, this function is assumed to store data
on the appropriate chip on the appropriate board, using some address lookup
table. Subroutine start_board initiates the calculation on board iboard, and
wait_board keeps polling the status flag of a board iboard until it finishes
the calculation. Finally, get_force retrieves the calculated force and potential
from the specified pipeline chip.

Figure 72 shows the time chart of the above implementation. In this figure,
it is obvious that there is some room for improvement.

Figure 73 shows the time chart of a more efficient implementation. We
can see that the force calculation pipelines are doing useful work for a larger
fraction of time.

Whether an optimization like this is actually useful or not must be quantitatively evaluated, and should be compared with other solutions. In the
following, we consider four alternatives:

(a) A naïve implementation described in Figure 72.
(b) A pipelined implementation in software described in Figure 73.
(c) A hardware solution with double buffering.
(d) A hardware solution with multiple independent I/O buses.

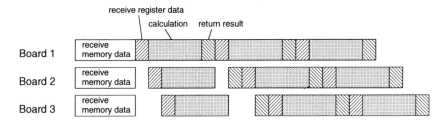

Figure 73   Parallel operation on a multi-board system with "pipelined" implementation.

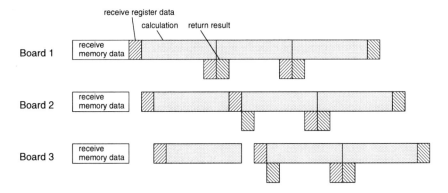

Figure 74   Parallel operation on a multi-board system with hardware double buffering.

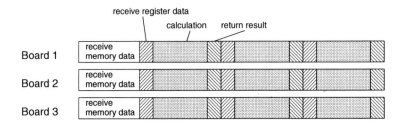

Figure 75   Parallel operation on a multi-board system with multiple independent I/O buses.

Figures 74 and 75 illustrates how options (c) and (d) work. Both require significant investment in hardware.

When we use option (a), the total computing time of a GRAPE system per one timestep (T_a) for the direct summation is expressed as

$$T_{\text{total,a}} = T_{\text{comm,a}} + T_{\text{force}} + T_{\text{misc}}, \tag{5.7}$$

where $T_{\text{comm,a}}$ is the time spent for data transfer between the host and GRAPE-3, T_{force} is the time taken to calculate the gravitational forces and potential for N particles, and T_{misc} is the time spent for miscellaneous computations on the host computer.

The time for communication (T_{comm1}) is

$$T_{\text{comm,a}} = N(t_{\text{mem}} + t_{\text{wreg}} + t_{\text{rreg}}), \tag{5.8}$$

where t_{mem} is the time taken to write the data of one particle to GRAPE memory, and t_{wreg} and t_{rreg} are the time to write or read the on-chip registers for one particle, respectively. For simplicity, we assume $t_{\text{mem}} = t_{\text{wreg}} = t_{\text{rreg}} = t_{\text{comm}}$.

The time for the force calculation on GRAPE-3 (T_{force}) is expressed as

$$T_{\text{force}} = \frac{N^2 t_{\text{pipe}}}{n_{\text{pipe}}}, \tag{5.9}$$

where t_{pipe} is the clock period of the pipeline chips and n_{pipe} is the total number of pipeline chips. In the following, we ignore T_{misc} for simplicity. For relatively simple time integration, T_{misc} is typically smaller than $T_{\text{comm,a}}$.

We can define the efficiency of the algorithm by

$$\eta = \frac{T_{\text{force}}}{T_{\text{total}}}, \tag{5.10}$$

since this η indicates the fraction of the time in which GRAPE hardware is doing useful work. The sustained performance is the theoretical peak multiplied by this efficiency factor.

For option (a), the efficiency is given by

$$\eta_a = \frac{1}{1 + b/N}, \tag{5.11}$$

where b is the ratio between the communication performance and calculation performance, defined as

$$b = \frac{3t_{\text{comm}} n_{\text{pipe}}}{t_{\text{pipe}}}. \tag{5.12}$$

If the communication is relatively fast, b is small. If calculation on GRAPE is fast, b becomes large.

The actual value of b varies widely among different GRAPE hardware and within the same GRAPE systems with different configurations, but is in between 10^3 and 10^5. Note that b has the same meaning as $n_{1/2}$ of vector processor [HJ81]. For $N = b$, the efficiency η is 0.5.

When we use option (b), the time per one timestep is expressed as

$$T_{\text{total,b}} = \max \left(\frac{2n_{\text{board}} - 2}{3n_{\text{board}}} T_{\text{comm,a}}, T_{\text{force}} \right) + \frac{n_{\text{board}} + 2}{3n_{\text{board}}} T_{\text{comm,a}}. \quad (5.13)$$

Here we neglected the loss at the beginning and end of calculation (see Figure 73). This treatment is alright if $N \gg n_{\text{pipe}}$. In this case, the efficiency is given by

$$\eta_{\text{b}} = \begin{cases} N/b, & (N < k_{\text{b}}b) \\ 1/[1 + (1 - k_{\text{b}})b/N)], & (N \geq k_{\text{b}}b) \end{cases} \quad (5.14)$$

where k_{b} is defined as

$$k_{\text{b}} = \frac{2n_{\text{board}} - 2}{3n_{\text{board}}}. \quad (5.15)$$

From a similar consideration, the efficiency of options (c) and (d) is given by

$$\eta_{\text{c}} = \begin{cases} N/b, & (N < 2b/3) \\ 1/[1 + b/3N)], & (N \geq 2b/3) \end{cases} \quad (5.16)$$

and

$$\eta_{\text{d}} = \frac{1}{1 + (1 - k_{\text{b}})b/N}. \quad (5.17)$$

Figure 76 shows the efficiency of these four options as a function of the normalized number of particles N/b, for the case of $n_{\text{board}} = 4$.

The improvement from option (a) to option (b) is fairly large, even though this improvement is achieved purely by software. On the other hand, the additional gain achieved by using options (c) or (d) is quite small. In fact, for $N < k_{\text{b}}b$, options (b) and (c) are the same, while for $N \geq k_{\text{b}}b$, options (b) and (d) are the same. When we take into account the additional complexity, in most cases it is not worthwhile implementing additional hardware for double buffering or for an additional host interface.

With the methods described above, the improvement in performance is fairly limited. Even for option (d) with a very large n_{board}, the maximum possible gain in efficiency is a factor of three. This is because the communication time is lower-bounded by the time taken to store particles to memory in the GRAPE part, which is independent of the number of interfaces.

By using a different parallelization strategy, we can improve the performance of option (d) further. In this strategy, we divide the boards into p groups with n_{boards}/p boards each. Each group calculates the force on all particles from its subset of particles. The host computer adds up p forces to obtain the total force on a particle. This is what is actually implemented on

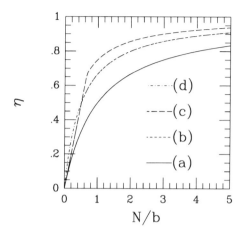

Figure 76   Relative efficiency of various multi-board implementations.

GRAPE-4, though the reason we used this strategy on GRAPE-4 is somewhat different from the efficiency consideration for simple direct summation.

With this scheme, the time for communication is given as

$$T_{\text{comm,d2}} = N[t_{\text{mem}}/p + (t_{\text{wreg}} + t_{\text{rreg}})p/n_{\text{board}}]. \tag{5.18}$$

The value of p for which this communication time is at a minimum is given by $p_{\text{opt}} = \sqrt{n_{\text{board}}/2}$. The time for communication decreases roughly as $1/\sqrt{p}$.

If the multiple GRAPE boards are connected by a network which is *faster* than the connection to the host computer, we could further improve the performance.

Here, we discussed mainly the performance of the direct summation method with a shared timestep. The techniques discussed here are, however, applicable to other methods, including the treecode and individual timestep scheme.

5.5.2   Concurrent operation of host and GRAPE

In the previous section, we ignored the calculation time on the host computer. This simplification is alright when we consider the performance of the direct-summation, shared-timestep algorithm. However, with more sophisticated algorithms, the time spent on the calculation on the host computer cannot be ignored. Here, we discuss the change in the algorithm which allows the calculation on the host and calculation on GRAPE to proceed concurrently, thereby decreasing the total execution time. Here, we ignore the communication time needed to simplify the discussion.

First, we consider the individual (hierarchical) timestep algorithm. For simplicity, we consider a system with one host and one board with a single pipeline processor. The generalization to a multiple-pipeline system is straightforward. One timestep of the hierarchical timestep algorithm on GRAPE consists of the following steps:

(a) Construct list of particles which share the same minimum time. Update the current time.
(b) For each particle in the list, predict its position at the current time on the host and send that data to GRAPE. This operation is necessary to avoid self interaction. (predict)
(c) For each particle in the list, calculate the force using GRAPE. (force)
(d) For each particle in the list, apply the corrector and update time and timestep. (correct)
(e) For each particle in the list, update the data on GRAPE. (update)
(f) Go back to step (a)

The advantage of the hierarchical timestep algorithm is that steps (b) through (e) can be parallelized over all particles in the present list. However, a naïve implementation of the above description is quite inefficient, since GRAPE works only in step (c). In all other steps, only the host computer is working.

Steps (b)–(e) can be performed in a parallelized fashion over particles because operations on different particles are mutually independent. Strictly speaking, step (e) is problematic, since it changes the result. However, for steps (b)–(d), we can overlap the execution. we can start the force calculation for a particle when its position is predicted. We do not have to wait for all particles to be predicted. To show a pseudocode, a naïve implementation would look as follows:

```
perform step(a)
do i=1, nlist
   call predict(list(i))
enddo
do i=1, nlist
   call start_grape(list(i))
   call wait_grape
   call get_force(list(i))
enddo
do i=1, nlist
   call correct(list(i))
enddo
do i=1, nlist
   call update(list(i))
enddo
```

The second loop shows schematically how the host computer uses GRAPE. The following implementation gives the same result up to a roundoff error (the difference between the predictor on GRAPE and predictor on the host might cause the difference in the roundoff error).

```
perform step(a)
do i = 1, nlist
    call predict(list(i))
    call start_grape(list(i))
    call wait_grape
    call get_force(list(i))
    call correct(list(i))
enddo
do i=1, nlist
    call update(list(i))
enddo
```

It is possible to change the ordering of prediction and other things so that the prediction of the next particle and correction of the previous particle are performed while GRAPE is calculating the force. This can be achieved by the following pseudocode:

```
perform step(a)
call predict(list(1))
do i = 1, nlist
    call start_grape(list(i))
    if (i .lt. nlist) call predict(list(i+1))
    if (i .gt. 1)     call correct(list(i-1))
    call wait_grape
    call get_force(list(i))
enddo
call correct(list(nlist))
do i=1, nlist
    call update(list(i))
enddo
```

If the communication time is negligible, this optimization could achieve a performance improvement of a factor of two at maximum, in the case in which the calculation time of the host and that of GRAPE are the same. With the present GRAPE-4, the improvement in the performance is limited, since the communication takes a significant time, as will be shown in Section 5.6.

The same technique can be used with a multiple-pipeline system, although the program becomes rather lengthy. A similar technique can also be used with the Barnes–Hut treecode. However, in the case of the treecode, we have to perform the force calculation on a block of particles at the same time as the tree traversal for the next block of particles. This is complicated to implement efficiently, unless the time for the interrupt handling and context switching is very small.

5.6   Performance

5.6.1   GRAPE-4

In this section, we present the measured speed of the GRAPE-4 system for the individual timestep algorithm.

To measure the performance, we performed systematic test runs on various hardware configurations. As the initial conditions, we used the King profiles with a non-dimensional central potential W_c of 3, 7 and 10. We changed the number of particles from 128 to 524 288, and measured the speed for the configurations with one, two and three clusters. The fourth cluster was unavailable at the time of the experiments.

The system of units is chosen so that the total mass of the system M and the gravitational constant G are both unity. The total energy of the system E is $-1/4$ [HM86a]. The softening parameter is $1/N$, where N is the number of particles. The mass of all particles is $m = 1/N$. We integrated the system for one time unit and measured the CPU time on the host. The host computer was a DEC Alpha AXP 3000/900 with 448MB of memory.

Figure 77 shows the calculation speed in Gflops for the runs from King models with $W_c = 3$. Triangles, squares and circles show the speed of 1-, 2- and 3-cluster configurations. Solid, short-dashed and long-dashed curves are the numbers obtained by the simple performance model, which will be discussed later. The speed achieved is roughly proportional to the number of particles for $10^3 < N < 10^5$, and almost independent of the number of clusters. In this range, the use of more than one cluster actually decreases the overall speed, since the total performance of the system is limited by the speed of the host computer. As described in Section 4.8, the amount of communication increases as the number of clusters increases. For 256K particle runs, two- and three-cluster configurations are actually faster than the one-cluster configuration. For 512k particle runs the three-cluster configuration is the fastest. The CPU time per time unit is about eight hours for a 512k particle run on a 3-cluster system. Thus, such a simulation [FM97a] is feasible, at least for the crossing time scale.

Figure 78 shows the speed of the 3-cluster configuration for various initial models. For small-N runs, the speed is lower for a higher central concentration. For large-N runs, the difference is very small. This difference is due to the difference in the average number of particles, n_s, to share the same time, which is shown in Figure 79. If this n_s is smaller than the number of force calculation pipelines, the performance of GRAPE-4 becomes lower, since some of the pipelines are not used. The number of pipelines is 94 in GRAPE-4.

Figure 79 shows that n_s is quite well approximated by $\sqrt{N}$, though it depends somewhat upon the structure of the cluster. This dependence is different from the theoretical model [Mak91c], which predicts

$$n_s \propto N^{2/3}. \tag{5.19}$$

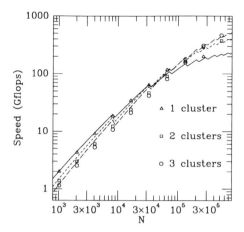

Figure 77   The measured performance of GRAPE-4 in Gflops and the
performance model.

The disagreement of the theoretical prediction and experimental result for
n_s is caused by the difference in the average number of timesteps s_{avg}. The
theoretical model is derived using the assumption that the average number of
timesteps per particle per time unit, s_{avg}, should satisfy

$$s_{avg} \propto N^{1/3}, \tag{5.20}$$

and the number of system timesteps per time unit (the time average of the
inverse of the minimum timestep), s_b is $O(N^{2/3})$. Figure 80 shows s_{avg} as the
function of the number of particles N. The behavior of s_{avg} is systematically
different from the theoretical prediction. The formula

$$s_{avg} \propto N^{1/6} \tag{5.21}$$

seems to be a good fit.

The experimental result of Equation (5.21) is a direct consequence of the
fact that we used the timestep criterion based on the time derivatives of the
acceleration. Any criterion which is expressed as a non-dimensional function
of the acceleration and its time derivatives shows the dependence of the form
of Equation (5.21) [MH88]. On the other hand, Equation (5.20) is derived
from the assumption that the timestep should be proportional to the average
interparticle distance. In other words, the timestep criterion we have been
using might fail to resolve close encounters for very large N. For the range of
N we tested, the difference between $N^{1/3}$ and $N^{1/6}$ is only a factor of four.
Therefore, it is unlikely that the close encounters are not resolved properly.

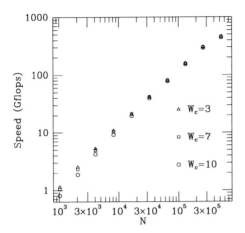

Figure 78   The dependence of the speed on the structre of the cluster.
Triangles, squares and circles denote the result for King models with $W_c = 3, 7$
and 10, respectively. (Reproduced from Makino *et al.* (1997).)

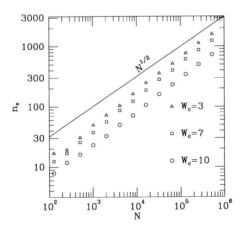

Figure 79   The average number of particles to share the same time n_s plotted
as a function of the number of particles in the system N. Symbols have the same
meaning as in the previous figure. (Reproduced from Makino *et al.* (1997).)

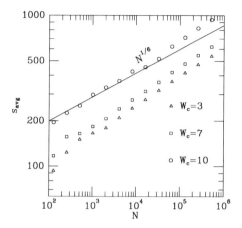

Figure 80   The average number of timesteps per unit time. (Reproduced from Makino *et al.* (1997).)

Conceptually, the time needed to integrate one particle on GRAPE-4 system is given as follows:

$$T_{\text{step}} = T_{\text{host}} + T_{\text{grape}} + T_{\text{comm}}, \tag{5.22}$$

where T_{host}, T_{grape} and T_{comm} are the time to integrate the orbit of one particle on host, the time to calculate the force on one particle on GRAPE-4, and the time to transfer data between the host and GRAPE-4, respectively. In reality, the actual time can be made shorter than this estimate, because the calculation on GRAPE and that on the host are partially overlapped.

The three terms on the right-hand side of Equation (5.22) are expressed as

$$T_{\text{host}} = 250 t_{\text{host}}, \tag{5.23}$$

$$T_{\text{grape}} = \gamma \frac{3 N t_{\text{pipe}}}{47 n_{\text{cl}} n_{\text{pr}}}, \tag{5.24}$$

$$T_{\text{comm}} = (34 + 20 \gamma n_{\text{cl}}) t_{\text{comm1}}. \tag{5.25}$$

Here, t_{host} is the average time taken for the host computer to perform one floating-point operation. For the Hermite scheme, the total number of floating-point operations per particle per timestep is $200 \sim 300$. The actual speed of the host computer for that part of the code executed on the host is quite low, like $15 \sim 20$ Mflops. We thus estimate that $t_{\text{host}} = 5 \times 10^{-8}$ sec. In Equation (5.24), N is the number of particles in the system, t_{pipe} is the cycle time of the HARP chip, and n_{cl} and n_{pr} denote the number of clusters and the number of processor boards per cluster, respectively. The parameter γ is the loss of parallel efficiency of multiple pipelines, which is estimated as

$$\gamma = \left[\frac{n_s + n_{vp} - 1}{n_{vp}} \right] \frac{n_{vp}}{n_s}, \tag{5.26}$$

here, $[x]$ denotes the maximum integer which does not exceed x, and n_{vp} is the number of virtual pipelines. In a real GRAPE-4 system, $n_{vp} = 94$. For the present GRAPE-4, the clock cycle of HARP chips is 32 MHz and the number of processor boards per cluster is 9, i.e., $t_{\text{pipe}} = 3.125 \times 10^{-8}$ sec and $n_{\text{pr}} = 9$. In Equation (5.25), t_{comm1} is the time taken to transfer one word (4 bytes) between the host and a control board. The number of words to be transferred between the host and a control board is $34 + 20n_{\text{cl}}$ in the present implementation of the control board.

For the communication time, we used the value $t_{\text{comm1}} = 1.8 \times 10^{-7}$ sec. The raw speed of the DMA on the TURBOchannel interface is close to 100 MB/s, which is translated to $t_{\text{comm1}} = 4 \times 10^{-6}$ sec. The actual time spent for the communication is significantly larger simply because the data copying within the main memory takes rather a long time, as discussed in Section 4.2. The DMA function of HIB can directly access the user address space. Even so, the user process needs to pack/unpack the data, because a single DMA operation can only transfer a block of memory.

We can see from Figure 77 that the theoretical performance model agrees well with the experimental results for both the small- and large-N, but tends to overestimate the speed for intermediate values of N. This is partly because our estimate for γ is too crude, and partly because the estimate for the communication time for multi-cluster configurations is too simplified.

The actual performance for a relatively small number of particles ($N < 10^5$) is noticeably lower than our original assessment [MTTS94]. This is mainly because we underestimated t_{comm1}. In our original assessment, we neglected the time needed to move data within the main memory.

6

Science by Special-Purpose Systems

6.1 Planetary formation

6.1.1 Runaway growth of protoplanets

The standard model for the formation of the planets is that they are formed from smaller planetesimals. Planetesimals themselves are formed through gravitational instability of the dust disk (for a recent review see Lissauer [Lis93]).

Planetesimals are gravitationally scattered by close encounters with others, and in some cases physically collide and coalesce. As a result, planetesimals slowly grow to finally form planets.

N-body simulation is the most direct tool to study such a process. To perform an N-body simulation of planetesimals, however, requires tremendous computer power. As a result, N-body simulation was not the main tool for study of the planetary accretion process in the 1970s and 1980s. Aarseth and his collaborators performed several impressive calculations, but important findings came mostly from analytic models and Fokker–Planck calculations. In the 1980s, however, the theory of planetary formation was confronted with a serious difficulty. It would take too much time to form planets from planetesimals. Mars and planets inside its orbits were formed easily, but Jupiter and the other outer planets should not, according to the standard theory, have been formed.

The "solution" to this difficulty was provided by Wetherill and Stewart [WS89], who re-evaluated the dependence of the growth rate of planetesimals on their masses taking into account the effect of the dynamical friction, in other words, energy equipartition.

In planetesimal dynamics, the orbits of particles are close to circular. The close encounter of two particles tends to generate the deviation from the circular orbit (see Ida [Ida90] for detailed numerical simulation of the effect of close encounters). As the result of many such encounters, the deviations of orbits from the circular orbits tend to randomize. Thus, we can discuss the evolution of the distribution of the random motion (deviation from the circular orbit) in much the same way as we discuss the random velocity in a normal statistical system.

The random velocity, in thermal equilibrium, would be equipartitioned over particles with different masses. In other words, the orbits of heavier planetesimals are, on average, closer to the circular orbit than those of lighter planetesimals.

The cross-section of the physical collision between planetesimals depends strongly on the distribution of the relative velocity through gravitational focusing. For high-velocity encounters, the impact parameter b and actual separation at the closest approach are essentially the same. However, in low-velocity collisions, the actual minimum separation is smaller than the impact parameter at infinity because of the mutual gravity. If the relative velocity is small, the cross-section becomes much larger.

The random velocity is smaller for heavier planetesimals because of the energy equipartition. Thus, on average, a collision cross-section is larger for massive planetesimals, and therefore massive planets grow faster than other planets. Stewart and Wetherill [SW88a] called this effect "runaway growth".

Ida and Makino [IM92a], [IM92b] investigated the evolution of the velocity distribution of planetesimals and that of a protoplanet embedded in a swarm of planetesimals. They confirmed that the equipartition is actually achieved, and therefore the runaway growth should take place. These calculations were performed on GRAPE-2. In these calculations, physical collision and coalescence of planetesimals are neglected, because their main goal was to investigate the evolution of the velocity distribution. In addition, the speed of GRAPE-2 was not quite sufficient to perform a simulation of the planetary growth.

Figure 81 shows the time evolution of the r.m.s. random velocity for 100 equal-mass planetesimals distributed uniformly. In planetesimal dynamics, it is convenient to express the random motion in terms of eccentricity and inclination of orbits. In addition, it is helpful to normalize them by means of the Hill radius, defined as

$$R_{\mathrm{H}} = \left(\frac{m}{3M_\odot} \right)^{1/3} a, \qquad (6.1)$$

where m is the mass of a planetesimal, $M_\odot$ is the mass of the Sun, and a is the distance between the planetesimal and the Sun. Thus, we plot normalized eccentricity and inclination $< e_{\mathrm{H}}^2 >^{1/2}$ and $< i_{\mathrm{H}}^2 >^{1/2}$. First the eccentricity rises quickly and then inclination follows. In the later stage, they evolve to follow the relation

$$< e_{\mathrm{H}}^2 >^{1/2} \sim 2 < i_{\mathrm{H}}^2 >^{1/2}. \qquad (6.2)$$

This behavior is in excellent agreement with the heating rate obtained by the three-body scattering experiment [Ida90], and theoretical explanation was given later by Ida et al. [IKM93].

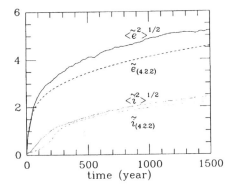

Figure 81 Evolution of the random velocity of planetesimals. (Reproduced from Ida and Makino (1992) by permission of Academic Press.)

Kokubo and Ida [KI96] performed simulation of the planetary accretion under the assumption of perfect accretion (any physical collision leads to coalescence) using GRAPE-4. Their initial condition is 3000 equal-mass planetesimals. After 20 000 orbits, the most massive particle becomes 300 times heavier than the initial mass, while the average mass of particles is increased only by a factor of two (see Figure 82).

Figure 83 shows the evolution of the mass distribution of planetesimals. Initially, all planetesimals have the same mass. The distribution quickly relaxes to a power-law form, expressed roughly as

$$n(m) \propto m^{-2.5}. \tag{6.3}$$

Makino *et al.* [MTES97] gave the theoretical mechanism of the formation of this power-law distribution. Their argument is summarized as follows. The rate of the collision between two planetesimals of masses m_1 and m_2 ($m_1 < m_2$) can be approximated as

$$P \simeq C m_1 m_2^{4/3} = C m_1^{\nu} g(m_2/m_1), \tag{6.4}$$

where C is a constant which depends upon the total surface mass density, ν is the dependence of the collision rate on the mass (here $\nu = 7/3$), and $g(x)$ is some arbitrary function of x. Equation (6.4) is obtained by taking into account the effect of the energy equipartition on the scattering cross-section and the volume density of planetesimals.

Tanaka *et al.* [TIN96] have shown that the stationary state distribution is given by

$$n(m) \propto m^{(\nu+3)/2}. \tag{6.5}$$

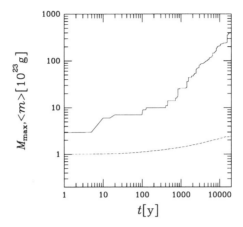

Figure 82   Time evolution of the mass of the most massive planetesimal (solid
curve) and the average mass (dashed curve). (Reproduced from Kokubo and Ida
(1996) by permission of Academic Press.)

Thus, by adopting $\nu = 7/3$, we obtain $n(m) \propto m^{-8/3}$, which is in good
agreement with the result of numerical simulations.

If this runaway growth can continue to the mass of actual planets, any planet
can be formed in a very short time. However, Ida and Makino [IM93] suggested
that the runaway growth would slow down when the protoplanet becomes
massive enough to affect the velocity dispersion of nearby planetesimals. They
showed, using both semi-analytical theory and N-body simulation, that the
heating by the protoplanet becomes significant if its mass is larger than that
of planetesimals by a factor of 100 or so, which corresponds to about 1/10 of
the mass of the terrestrial planets.

6.1.2   Orbital repulsion

The timescale of the formation of the planet is of course an interesting problem, but the central problem of the theory of planetary formation is the origin
of the separation between planets. Everyone has heard of the so-called Bode's
law, which mysteriously fits the separation between planets.

Traditional analytic theories give no clue to the separation between the
planets, because they treat the distribution of planetesimals and protoplanets
as spatially homogeneous. This assumption is entirely wrong when we are
interested in the late stage of planetary accretion. Thus, the only practical
way to investigate the separation of the planets is to perform an N-body
simulation of their formation process.

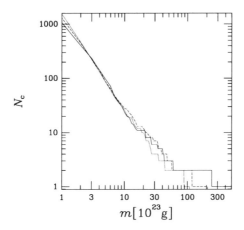

$$m[10^{23}\text{g}]$$

Figure 83   Time evolution of the mass distribution of a planetesimal.
(Reproduced from Kokubo and Ida (1996) by permission of Academic Press.)

Kokubo and Ida [KI95] performed a simulation of the system of two protoplanets and many planetesimals using HARP-2 (a smaller prototype of
GRAPE-4), and found a very interesting result – that the separation between
two planets tends to grow to around 5 r_H, where r_H is the Hill radius of the
protoplanets. They also give a qualitative explanation for this behavior. They
called this effect the "orbital repulsion".

When the radial separation of two protoplanets is less than $5r_\mathrm{H}$, they tend
to be scattered at close approach, and both orbits become eccentric. In an
isolated environment, these eccentric orbits are stable. In other words, there
will be no secular evolution of orbital elements. However, since there are still
many planetesimals around, the orbits of protoplanets are circularized by the
dynamical friction. Thus, protoplanets are again forced into unstable circular orbits, and then scattered by close encounter. Thus, the overall process
effectively increases the radial separation.

Kokubo and Ida [KI97] performed the simulation of planetary formation
using the same calculation code as used earlier [KI96], but for a wider spatial
range and a larger number of particles. They showed that several protoplanets
are formed and they grow while keeping the mutual separation to be around
$5 \sim 10$ r_H. The separations between the outer planets are actually close to
this separation, and their result very strongly suggests that the orbital repulsion determines the separation between outer planets. The present separation
between inner planets is significantly larger, and seems to require a different
explanation.

6.1.3   Algorithm and implementation

In this section, we overview the algorithms used for planetary simulation. The basic integration scheme used in most N-body studies discussed in previous sections is a very straightforward implementation of the Hermite scheme described in Chapter 5, with a simple procedure to implement collision and accretion. The only difference from the basic scheme is that the solar gravity is calculated on the host computer to improve accuracy. Since the solar gravity is typically many orders of magnitude larger than the mutual gravity of planetesimals, it significantly reduces the roundoff error without increasing the total calculation cost.

Traditionally, many researchers have proposed more complex algorithms, and some have been widely used. In the following, we first discuss the collision detection, and then discuss the pros and cons of more sophisticated treatments.

Collision detection and implementation of collision

In most existing studies, the planetesimals and protoplanets are assumed to be spherical. Thus, collision occurs if the distance between two particles is smaller than the sum of the radii of two particles. When we calculate the force on one particle, we calculate the distance to all other particles. Thus, the detection of the collision is quite simple.

With GRAPE-4, in principle, the implementation is simple, since GRAPE-4 can construct the list of particles within a given distance from any one particle. If we set this radius as twice the radius of the particle itself, we can detect all possible collisions.

The present implementation [KI97] does not make use of this hardware for the neighbor search because of several technical problems with the neighbor list of GRAPE-4. Instead, it first checks the timestep of particles. If the timestep is smaller than a critical value, it implies that a close neighbor might be there. The actual nearest neighbor is searched only for this case. In the current implementation, the search is actually performed on the host computer, since the calculation cost of the search is quite small.

There are a number of possible ways to reduce the cost of the neighbor search in the case of planetary simulation. The simplest is to use the standard linked-list scheme described in Chapter 3. We could further reduce the cost of reassigning particles by using a polar grid and possibly in a corotating reference frame.

In practice, we could minimize the cost of the neighbor search by having the resolution of the cell smaller than the average interparticle distance, since we are dealing with a relatively small number of particles, and because the radius of the particles is smaller than the average interparticle distance by several orders of magnitude.

When two particles physically collide, we simply form a new particle with the mass and momentum calculated as the sum of two particles. A technical problem here arises from the fact that we are using the individual timestep scheme, and therefore when a particle detects the collision, the other particle might not have the same time.

To avoid the complexity, we simply synchronize all particles when collision is detected, by forcing all particles to be integrated to the time of collision. This does not significantly increase the total calculation cost, since the maximum number of collisions is only N, while the total number timesteps is much larger than N, at least currently.

In future, with faster machines, we will be able to use a much larger number of particles. In this case, we cannot synchronize all particles at each collision. We need to limit the particles to be synchronized to be the neighbors of collided particles.

Refinement of the time integration algorithm I. Special treatment of the solar gravity

The planetary simulation requires a very high accuracy because it covers many orbital periods, up to 10^6 or even longer. The change of the orbital elements of each particle due to the numerical error should be small enough to guarantee the validity of the result. Thus, if we require the final energy conservation of, say, something like 10^{-5}, we have to conserve the energy to better than 10^{-11} per orbit. In the case of a standard Hermite scheme, this implies that we need more than 10^3 timesteps per orbit, which means a total number of timesteps per particle of over one billion.

When we consider the nature of the system, however, there is an obvious way to improve the efficiency. The solar gravity is the dominant force. Therefore, the actual orbits of the planetesimals are quite close to the Kepler orbit. We can greatly improve the accuracy if we integrate the deviation from the Kepler orbit, instead of integrating the orbit directly.

This is essentially the same idea as the classical Encke's method to integrate the orbits of planets and comets under the effect of perturbations from other planets. The basic idea is the following. Consider the case where we have the equation of motion written as

$$d^2x/dt^2 = \mathbf{F}(\mathbf{x}) + \mathbf{f}(\mathbf{x}), \tag{6.6}$$

with the condition $|\mathbf{F}| \gg |\mathbf{f}|$, and we know the analytic solution of the equation

$$d^2x/dt^2 = \mathbf{F}(\mathbf{x}), \tag{6.7}$$

as $\mathbf{x} = \mathbf{x}_0(t)$. We can rewrite Equation (6.6) as

$$d^2\Delta\mathbf{x}/dt^2 = \mathbf{F}(\mathbf{x}) - \mathbf{F}(\mathbf{x}_0) + \mathbf{f}(\mathbf{x}), \tag{6.8}$$

where $\Delta\mathbf{x}$ is defined as

$$\Delta\mathbf{x} = \mathbf{x} - \mathbf{x}_0. \tag{6.9}$$

It should be noted that the transformation from Equations (6.6) to (6.8) is exact, with no approximation involved.

The advantage of using Equation (6.8) instead of the original equation is that $\Delta\mathbf{x}$ is several orders of magnitude smaller than $\mathbf{x}$, and therefore the truncation error of the time integration is reduced by roughly the same factor. To achieve the same accuracy, we can use a significantly longer timestep with this method.

Ida and Makino [IM92a] implemented the above method in cylindrical coordinates, and used polynomial approximation (8th order) to solve the Kepler equation. They used a classical four-stage Runge–Kutta scheme with a shared timestep to integrate the derived equation of motion. If the random motion of particles was small and close encounters were rare, they could use a timestep as large as $1/10$ of the orbital period and still retain very high accuracy. However, because they did not implement the individual timestep algorithm, the close encounters were handled by reducing the timestep for the total system, thus resulting in a rather quick increase in the total calculation cost when random motion becomes larger.

A serious disadvantage of this method is that the accuracy can actually be worse than that of the simple direct method when applied to close encounters of planetesimals, where the mutual gravity is much larger than that of the solar gravity. Thus, some method which seamlessly combines the direct method and this method is necessary, but has not yet been devised.

Emori *et al.* [EIN93] describe this method in the Hill local coordinate system in some detail.

Refinement of the time integration algorithm II. Symplectic or symmetric methods

Recent advances in research into the symplectic integration scheme have demonstrated that symplectic methods can dramatically improve the accuracy of the long-term integration of Hamiltonian systems [SSC94]. Many researchers chose the two-body Kepler problem to demonstrate the superiority of the symplectic method over traditional schemes. For the long-term integration of planets, variations of the symplectic method are now widely used [WH91], [ST92], [ST94], [LD94]. Thus, planetesimal dynamics looks like the obvious application.

However, symplectic schemes have not been used in this field because of several theoretical difficulties. First, as described in Sanz-Serna and Calvo [SSC94], symplectic schemes lose their advantage over traditional schemes when a variable timestep is used. Saha and Tremaine [ST94] described one method of using different timesteps for different particles. However, in their method it was not practical to *change* the stepsize assigned to each particle. Thus, their method would be useful for the present solar system, in which planets are well separated from each other. However, for the system of planetesimals, an individual and variable timestep is necessary.

Hut *et al.* [HMM95] solved the first problem, though not really for the symplectic schemes. It has been known that the time-symmetric schemes have good properties similar to those of the symplectic schemes, at least for Hamiltonian systems with periodic solution. For symmetric schemes, they introduced the concept of the timestep symmetrization, which allows the timestep to be exactly the same when the time reversal is performed.

In the standard method of automatic timestep adjustment, the size of the timestep is determined using the estimate of the error of the previous step. Thus, if the timestep is shrinking, we systematically overestimate the new timestep, and if the timestep is enlarging, we systematically underestimate the timestep.

As a result, the deviation from the true orbit is systematically different when we do the time reversal. Therefore, the system cannot go back to the original initial condition when time reversal is applied. In the case of the constant timestep, we naturally go back to the original state, because the scheme is time-symmetric.

To force the time reversal to work with the variable timestep, Hut *et al.* [HMM95] introduced an implicit determination of the timestep based on the instantaneous estimate of the integration error. If this implicit error estimate is also time-symmetric, we can guarantee that the calculated stepsize would be exactly the same for the forward and backward path, and therefore the time reversibility is maintained.

Unfortunately, even with this timestep-symmetrization, to implement the symmetric algorithm with an individual timestep is still nontrivial. The problem with the individual timestep method is that we can use only the predicted positions of the particles not in the present blockstep. Thus, the interactions with particles not in the present blockstep cannot be symmetrized in time, unless their positions in future are already available.

In theory, we can construct a scheme which would actually try to use the future position, by performing something like the relaxation method in space-time. However, whether such a scheme would have any practical advantage remains to be seen.

In the case of the planetesimal dynamics, however, we can effectively achieve approximate global time-symmetricity easily. As described in the previous section, the solar gravity is the dominant term for most of the time, and the orbits of particles are very close to the circular Kepler orbit, with the eccentricity rarely exceeding 0.1. In other words, we might achieve significant improvement in accuracy by applying the time symmetrization only to the solar gravity, and the timesteps of particles are almost always constant. The timestep would divert from the constant value only during close encounters, which are relatively rare events.

Thus, if we use a time-symmetric integrator for individual particles, we might be able to improve the long-term behavior. The fourth-order Hermite scheme is time symmetric, when used with a full corrector (correct-to-convergence algorithm) instead of applying the corrector only for a fixed number of times. In practice, iteration of a few times is more than enough to effectively establish the time symmetry.

Kokubo and Makino [KYM97] implemented this full corrector scheme and made a detailed comparison with a standard Hermite scheme in PEC mode. Their conclusion is that the $P(EC)^2$ scheme actually improves the accuracy. The total energy is conserved with an accuracy several orders of magnitude better than that achieved with the standard PEC scheme. One could further improve the performance by not calculating the solar gravity after the first iteration. The mutual gravity is much smaller than the solar gravity, therefore we can ignore its change due to the correction in the position.

Reduction in the force calculation cost I. Aarseth's longitudinal subgrid method

Aarseth and his collaborators performed several direct N-body simulations of the planet formation process in the 1970s. The speed of the computer at that time was about $1/10\,000$ of what is available now. Thus, even with a few hundred particles, direct simulation was far too costly.

In the case of the planetary system, we can ignore the interactions with all particles other than very close neighbours. In addition, since particles are distributed in a relatively thin ring (well, if possible, they would simulate a system-wide radial range; however, with the limited computing power available, it was impractical), the linked list method with a one-dimensional grid in the azimuthal direction worked fine.

A fundamental problem with this scheme is that the total force from "neglected" particles is not actually negligible. Moreover, the force changes discontinuously when one particle enters or goes out of the interaction region.

Reduction in the force calculation cost II. The sliding box method

Another way to reduce the calculation cost is to limit the region even further, by cutting out a patch of the disk of planetesimals and simulating only that region. To take into account the effect of the particles outside the patch, we can apply the periodic boundary condition. To apply the periodic boundary, we use the Hill approximation, in which we assume that the deviation from the circular motion is small. In the Hill approximation, the coordinates of a particle are described by the deviation from the circular orbit at a certain radius. The center of the patch is used as the reference frame. The force on a particle in the patch is calculated as the sum of the forces from particles and its images in the eight surrounding patches.

Patches at the same distance from the sun do not move relative to the original patch. However, inner patches have a larger angular velocity, while outer patches have a smaller angular velocity. Thus, they move with constant velocity.

In the sliding box method, this motion of the inner and outer patches is taken into account by actually shifting them during simulation, which is why this method is called the sliding box method. When one cell goes too far, it is removed and a new patch is generated.

This method was invented by Wisdom and Tremaine [WT88] for studying the rings of Saturn. It has the major drawback that it cannot handle the inhomogeneity of the system, because of the very assumption of the periodic boundary. In addition, recent studies have revealed that this method is prone to the formation of spurious structures. Thus, the reliability of the result obtained by this technique is questionable. Reportedly, several people are now working on a code which puts the periodic boundary only in the radial direction, hoping to circumvent some of the problems caused by the box geometry. Whether or not this effort will be fruitful remains to be seen.

Richardson [Ric94], [Ric93] developed a quad-tree based algorithm for the sliding box method. Since the planetary system is thin, it sounds reasonable to use a two-dimensional tree. He implemented this scheme with an individual timestep, high accuracy force calculation and local tree reconstruction. This scheme would be quite useful to perform global calculations without using a periodic boundary.

6.1.4 Future

So far, the study of planetary formation with GRAPE has been extremely successful, being practically the first full three-dimensional simulation without a periodic boundary.

There are several future research directions. The first is to extend the calculation to increase the radial range and follow the later stage of evolution. Whether GRAPE will be really useful for simulation of the late stage is somewhat questionable, since the number of planets must eventually be reduced to around 10. However, if the effect of the remaining planetesimals as the source of the mass and dynamical friction is important, we might still need a very large number of particles. Also, it would be important to include the effect of gas.

The other research direction is the study of the planetary rings. The rings are different from the planetesimals in the sense that the physical collision between particles is much more frequent in the ring. So far, most of the numerical studies of the ring have adopted the "sliding box" method described above. However, whether such a method can give useful information on the dynamics of rings is not clear, and simulation of the ring without the assumption of the periodic boundary is the most direct and reliable way in which to test the reliability of these assumptions.

6.2   Star clusters

Star clusters have been regarded as the archetype of the self-gravitating N-body system, in the sense that it can be modeled reasonably well as the a collection of point-mass particles bound with self-gravity, and effects other than self-gravity such as the collision between stars, the effect of the tidal field from the parent galaxy, the effect of the evolution of stars, and the possible effect of primordial binaries can all be ignored. In this section, we first overview the theory of this idealized version of a globular cluster, and then proceed to more "realistic" models with non-gravity effects such as stellar evolution.

6.2.1   Idealized models

An ideal (from the point of view of the stellar dynamicists) globular cluster consists of point-mass particles all of the same mass. It exists in total isolation and has zero total angular momentum. In other words, it is essentially a collection of N point-mass particles distributed in space. The evolution of such a simple and idealized system has been the center of the theoretical and numerical study of the dynamics of star clusters during the last half of the century, and is not fully understood yet.

Violent relaxation

Consider an N-body system with a total mass M. In the limit of $N \to \infty$ (continuous limit), the state of the system is expressed by the mass distribution function in the six-dimensional phase space $f(\mathbf{x}, \mathbf{v})$. The evolutionary equation of the distribution function is expressed as

$$\frac{\partial f}{\partial t} + \mathbf{v} \cdot \nabla f - \nabla \phi \cdot \frac{\partial f}{\partial \mathbf{v}} = 0, \tag{6.10}$$

where ϕ is the gravitational potential given as the solution of the Poisson equation

$$\nabla^2 \phi = -4\pi G \rho. \tag{6.11}$$

Here, G is the gravitational constant and ρ is the volume density defined as

$$\rho = \int d\mathbf{v} f, \tag{6.12}$$

where integration is over the entire velocity space. Equation (6.10) is usually called the collisionless Boltzmann equation.

The system is in the dynamical equilibrium if $\partial f/\partial t = 0$. In this equilibrium state, both the distribution function and the potential field are time independent. As a result, each particle in the system conserves its energy and other integrals of motion (if they exist).

If the system is not in dynamical equilibrium, it will change the structure in a dynamical timescale, which is of the order of the freefall time, defined as

$$t_{ff} = \sqrt{G\rho}, \tag{6.13}$$

and eventually reaches an equilibrium.

Lynden-Bell [Lyn67] called this process "violent relaxation". In this process, particles in the system can change their energy and angular momentum, since the potential field is changing. In the case of the energy, the time change is expressed simply as

$$\frac{dE}{dt} = \frac{\partial \phi}{\partial t}. \tag{6.14}$$

Lynden-Bell thought that this effect of the time-varying potential would lead the system to some kind of statistical equilibrium, which he called the Lynden-Bell statistics. Whether his idea was really true or not will be addressed in Section 6.4.2. In any case, the system will reach a dynamical equilibrium.

Gravothermal catastrophe

If the number of particles is large enough, the system which reached the dynamical equilibrium would not evolve further, and the remaining part of this section is not necessary. Real stellar systems have only a finite number of particles, and therefore dynamical equilibrium is not the end of the story.

The potential field changes in time, since the number of particles is finite and each particle orbits in the system. As a result, particles change their orbits. To some extent, the effect of the change in the potential field can be understood as the statistical fluctuation in the potential field. However, since the particles which make the potential field are the same particles which are affected, the back reaction of the particles to the field cannot be ignored. Thus, the best way to understand this process is to treat the effects as the cumulative effect of two-body encounters [Cha43], [Spi96].

When regarded as the collection of two-body encounters, this effect of finite N works in the way similar to that of collisions of molecules in gas. Through these two-body encounters, the particles thermally relax and the system approaches the thermal equilibrium.

However, the thermal relaxation process of a self-gravitating N-body system is different from that of a gaseous system in two ways. First, the "mean free path" of particles in an N-body system is usually much larger than the system size, while the mean free path of the molecules in a gaseous system is a small fraction of the system size.

In a self-gravitating system, the "mean free path" is a somewhat unclear concept, since a particle is, in a sense, always having gravitational encounters with all other particles in the system. However, the effect of one particle is small.

Usually, we define the thermal relaxation time of the system as the time in which the cumulative change of either the velocity or the energy of a typical particle becomes order unity. Strictly speaking, there are several different definitions of the relaxation time, but all of them are of the order of

$$t_r = \frac{0.065v^3}{nm^2G^2\ln\Lambda},\tag{6.15}$$

where v, n and m are the r.m.s. velocity, number density and the mass of particles, respectively. The factor $\ln\Lambda$ is the Coulomb logarithm, which is approximated as $\ln(0.4N)$ for self-gravitating systems [LSH71]. The relaxation time is inversely proportional to n, the number density of field particles, since the number of encounters per unit time is directly proportional to n. It is proportional to third power of the velocity dispersion v, because the cross-section of the encounter with the same relative change in the velocity is proportional to v^{-4}. Finally, it is inversely proportional to the square of particle mass m, because the cross-section is proportional to m^2. For a given deflection angle, the impact parameter is proportional to the mass.

The ratio between the dynamical timescale t_{ff} and the "thermal" timescale t_r is given roughly as

$$\frac{t_r}{t_{ff}} = C\frac{N}{\ln(0.4N)},\tag{6.16}$$

where C is the constant of the order unity. We used Equation (6.13) and the relations $v \sim R/t_{ff}$ and $N \sim nR^3$. For star clusters, the relaxation time is several orders of magnitude larger than the freefall time.

In the gas dynamics, the fact that the mean-free path is smaller than the system size is the reason why we can apply the assumption of the Local Thermal Equilibrium (LTE). In other words, in the gravitational N-body system, the very assumption of the local thermal equilibrium, which makes it possible to use thermodynamical concepts like temperature or pressure, is totally broken.

To make the matter even more complex, a self-gravitating system (either made of point-mass particles or gas) has no thermal equilibrium.

Let us first consider an N-body system in isolation. We can assume that the system is in dynamical equilibrium. If the system is also in thermal equilibrium, the velocity distribution function must be the Maxwellian of the same temperature everywhere. This Maxwellian velocity profile, however, cannot be realized in a self-gravitating N-body system, since the density distribution should have a spherically symmetric halo with a density $\rho \propto 1/r^2$, which implies that the total mass is infinite.

Thus, any N-body system with a finite mass is not in thermal equilibrium. As a result of the thermal relaxation, they try to approach the isothermal, Maxwellian distribution function. In other words, the high-velocity tail of the distribution function, which is non-existant because only particles with negative energy are bound to the system, is constantly generated. These particles will naturally escape from the system. Thus, an N-body system spontaneously creates escaping particles.

This evolution of the system through the formation of escaping particles is theoretically important. However, in real star clusters the main driving force of the evolution is the structure formation within the cluster. The reason is that the local relaxation time is proportional to the density. Thus, the central high-density region can evolve much faster than the rest of the system. To give a clear understanding of the system, it will be helpful to consider the system in a spherical adiabatic wall [Ant62], [LW68], [HS78].

When we use an adiabatic wall, we can actually construct the isothermal distribution function. If we fix the total mass of the system and the radius of the wall, the equilibrium state is defined by a single parameter, for example, the temperature. It is, however, more convenient to parameterize the models by the contrast D of the density at the center of the system and that at the wall. Small D implies a high temperature. In particular, in the limit of the infinite temperature, D approaches unity. Large D implies a lower temperature.

Antonov [Ant62] found that the system has neutral stability at $D = 709$. Hachisu and Sugimoto [HS78] performed a stability analysis of the system by means of the perturbation equation, and found that the system is unstable against the redistribution of the heat if $D > 709$. In other words, if $D > 709$, the isothermal equilibrium state is unstable and would spontaneously develop a spatial structure if there is any small perturbation.

Hachisu et al. [HN78] performed a simulation of the evolution of the self-gravitating gas system with several different forms of heat conductivity. They found that if the heat conductivity is such that the central thermal timescale is shorter than that at the outside region, the evolution of the system would become self-similar. In other words, the central region continues to shrink, leaving the power-law halo behind it. The mass of the core decreases in time and the central density reaches infinity within a finite time.

In the following, we follow the description by Lynden-Bell and Eggleton [LE80]. Formally, a self-similar solution for a physical quantity y is expressed as

$$y(r,t) = y_0(t)y_*[r/r_0(t)]. \tag{6.17}$$

We can set $y_*(0) = 1$ without loss of generality. We can assume that in the limit of $r \to \infty$ there is no evolution, since the thermal timescale is longer at the outskirts. We can further assume that functions r_0 and y_0 are powers of the time t, since otherwise the self-similar solution cannot be constructed. Thus, if we express

$$r_0 = (t_0 - t)^\beta, \tag{6.18}$$

and

$$y_0 = (t_0 - t)^\gamma, \tag{6.19}$$

we have

$$y_0 = r_0^{\gamma/\beta}. \tag{6.20}$$

The ratio between the gravitational binding energy of the core and the thermal energy of the core would be constant. Therefore, we have

$$\sigma^2 \propto \frac{GM_c}{r_c} \sim \rho_0 r_0^2. \tag{6.21}$$

If we express ρ_0 as

$$\rho_0 = r_0^\alpha, \tag{6.22}$$

we have

$$r_0 = (t_0 - t)^{2/(6+\alpha)}. \tag{6.23}$$

Lynden-Bell and Eggleton [LE80] numerically obtained the self-similar solution for the gaseous model with heat conductivity which they believed would mimic the radial energy transfer in an N-body system, and found that the self-similar solution has the characteristic power law index of

$$\rho = r^{-2.21}. \tag{6.24}$$

The gravothermal catastrophe described above is the concept which, strictly speaking, can only be applied to gas systems. However, this concept has been strikingly successful in providing qualitative (and in many cases, quantitative) descriptions of the behavior of more sophisticated models without the assumption of LTE, such as the direct N-body calculation, and orbit-averaged Monte-Carlo and direct Fokker–Planck calculations.

Henon [Hen71] demonstrated that the N-body system would exhibit "core collapse" using Monte-Carlo calculation with 1000 shells. N-body simulation did show some collapse-like behavior, but it was difficult to see whether the collapse is really self-similar or not because of the limitation in the number of particles. The most beautiful demonstration of the self-similar nature of the collapse is by Cohn [Coh80], who is the first to use the direct integration of the Fokker–Planck equation to the study of the thermal evolution of the globular clusters. He found the power index to be -2.23, which is strikingly close to the value obtained by the gas model calculation of Lynden-Bell and Eggleton [LE80].

In the study of the gravothermal catastrophe and self-similar solution, N-body simulation did not play a major role. This is partly because more approximate methods, such as the gaseous models and Fokker–Planck calculations, gave reasonable results, and partly because the computer power available was quite limited even in the 1980s.

Gravothermal oscillations

The gravothermal catastrophe leads the central density of a star cluster to reach infinity in a finite time. What would happen after that? This question is of practical significance because a number of globular clusters appear to have undergone the collapse [CH84], [HD92]. Djorgovski and King [DK86] showed that 15% of galactic globular clusters have unresolved density cusps. Recent observations with the Hubble Space Telescope (HST) have demonstrated that some of these clusters have core sizes smaller than 0.03 arcsec, which is as far as one can go with present techniques [SK95], [YGBS94].

The theoretical study of evolution after the collapse was pioneered by Henon [Hen75], who incorporated the energy production by binaries into his Monte-Carlo calculation. He found that the whole cluster expanded homologously in the thermal timescale. Similar results were obtained by gaseous models and Fokker–Planck calculations.

When the central density becomes very high, the cross-section of the three-body close encounter, which causes the formation of a bound binary, becomes non-negligible. The binding energy of the binary is transferred to the kinetic energy of the third particle and the center-of-mass motion of the binary. This is essentially the same as the nuclear fusion reaction.

This energy production by binaries halts the collapse. The collapse of the core is driven by the outward heat flux, thus if a sufficient amount of energy is generated in the core, the collapse can be halted. This is the same as the evolution of a normal star, where the energy production from the nuclear fusion reaction balances the outward heat flux to form a main-sequence star in the stationary state.

For star clusters, one can construct such a stationary state solution. Goodman [Goo84] and Heggie [Heg84] obtained such a stationary solution using gas models.

Sugimoto and Bettwieser [SB83] found that the post-collapse expansion is unstable if the energy production rate is small. In the unstable case, the central density showed an oscillation with a large amplitude. They called this oscillation "gravothermal oscillation".

They modeled the post-collapse evolution of globular clusters using a conducting gas sphere with artificial energy production. The efficiency of the energy production is related to the total number of stars N, and for large N efficiency is small. Thus, their result implies that an N-body system with a number of stars larger than a critical number should show gravothermal oscillation.

Since several other calculations based on similar models did not reproduce the result of Sugimoto and Bettwieser [SB83], the validity of their result had been controversial for a few years after their first paper. However, calculations with improved accuracy confirmed that the oscillation takes place in both the gaseous models and Fokker–Planck (FP) calculations [Goo87], [HR89], [CHW89]. In particular, Goodman [Goo87] analyzed the stability of the stationary expanding solution, and found that it is unstable if N is large. Thus, it has now become well established that an idealized model of a globular cluster does exhibit gravothermal oscillation.

Whether such an oscillation would actually take place in a real cluster or not is, however, a rather different problem. Even if we consider an idealized system made of point-mass stars, the energy generation from the binary is no longer a continuous function of the local density, and there are only 20–30 stars in the core when the core becomes small enough so that the energy production from binaries can halt the collapse. Thus, only with N-body simulation can we convince ourselves that the oscillation actually takes place.

In the last 10 years, gravothermal oscillation has been one of the principal goals of N-body dynamicists. However, it was a rather distant goal, because the calculation cost of simulating the thermal evolution of the globular cluster increases as N^3. Hut et $al.$ [HMM88] estimated that the calculation cost of a 64k-body simulation is around 100 Gflops·year. So far, the largest simulation performed on conventional supercomputers is a 10 000-body simulation [SA96], which is, however, too short to determine whether the oscillation took place or not. The calculation was performed in part on a Cray YMP and in part on a C-90, and took about two CPU months.

Makino [Mak96] performed a simulation of the evolution of the system of point-mass particles with the number of particles being up to 32 768. The oscillation is clearly visible in simulations with large N. More importantly, the trajectory of the oscillation plotted in the plane of central density-central temperature indicated that the oscillation is of the same nature as that in gaseous models. The expansion of the core is supported by the heat flux from the outskirts, whose temperature is higher than that of the core. Figure 84 shows the time variation of the central density for simulations with 2–32k particles. The time is scaled so that the initial thermal relaxation time is the same for all runs. The curves are shifted vertically.

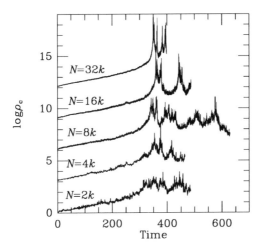

Figure 84   The time variation of the central density of simulated globular
clusters. (Reproduced from Makino (1996).)

The 32k particle calculation took about two months on one cluster of
GRAPE-4 running at a speed of 50 Gflops.

This calculation is the first clear demonstration that gravothermal oscillation actually occurs in an N-body system. As described in Section 4.1,
whether or not gravothermal oscillation would occur in an N-body system
had been the central problem in the study of the dynamical evolution of globular clusters, and also one of the main goals for the GRAPE project. We now
understand how an idealized globular cluster evolves, or, in other words, how
the collection of point-mass particles in isolation evolves.

6.2.2   More realistic models

The evolution of real globular clusters is much more complicated than that of
an idealized cluster which consists of equal-mass point particles.

Stars have different masses. Heavy stars tend to sink towards the center of
the cluster through the dynamical friction. Each star in the cluster evolves.
Massive stars would go type II supernova and become neutron stars. They are
likely to get a velocity kick and be ejected from the cluster altogether. Less
massive stars would evolve through the giant branch and end up as white
dwarfs.

Stars have finite sizes. Direct collision of two stars is quite rare, but their
dynamical effect cannot be ignored. Even when two stars do not physically
collide, a close encounter of two stars would cause the non-radial oscillation of

two stars, which implies that the encounter is inelastic. Thus, some sufficiently close encounters would result in tight binaries (tidal capture binaries). The physical radius of the stars in the giant branch is orders of magnitudes larger than that of the main-sequence stars. Thus, the effect of the inelastic collision might be quite important for giants.

Recent observations have revealed that the fraction of stars in globular clusters which are components of binary star systems is actually much higher than previous estimates. The classic work by Gunn and Griffin [GG79] concluded that the binaries were, if they existed, very rare in globular clusters. However, advances in the observation techniques in the 1980s resulted in a completely new view: the fraction of the binaries present in globular clusters might be as high as 10%. The binary fraction at the formation of the cluster might be even higher, since many binaries might have been disrupted by the close encounter with another star. Thus, globular clusters might be dominated by a primordial binary population, which changes the evolution after the core collapses almost completely [GH89].

McMillan *et al.* [MHM90], [MHM91] and Heggie and Aarseth [HA92] performed direct N-body simulation of globular clusters with primordial binaries. However, the number of particles is even more limited than that of the simulations without primordial binaries, because of the increase in the calculation cost. Theoretically, the increase in the calculation cost due to the existence of the binaries is smaller for larger N, so large-N simulations might be feasible with GRAPE-4. So far, because of the limitations in the software, we have not yet done a large simulation with primordial binaries.

The tidal field of the parent galaxy also plays an important role. Most studies of the tidal field assumed a time-independent tidal field. However, Weinberg [Wei94] and Gnedin and Ostriker [GO97] suggested that the effect of the time-dependent tidal field is much more important than previously speculated. All of these effects are easily investigated with N-body simulation, but very difficult or impossible to study with more approximate methods like Fokker–Planck calculation or gaseous models.

To show some examples of the work in this direction, Fukushige and Heggie [FH95] performed N-body simulations of the evolution of clusters with stellar-evolution and time-independent tidal force, using a GRAPE-3 system installed at Edinburgh University. They used essentially the same set of assumptions as in the Fokker–Planck calculation by Chernoff and Weinberg [CW90], except for the scaling of the timescales.

The basic assumption of the Fokker–Planck calculation is that the dynamical timescale is much shorter than any other evolutionary timescale. Thus, the distribution function is assumed to change adiabatically (see, for instance, Spitzer [Spi96]). This assumption is fine for an idealized cluster with no stellar evolution or galactic tidal field, since the only timescales we have to take care of are the dynamical timescale and relaxation timescale, which are in fact well separated in real globular clusters.

However, once we start to construct more realistic models of globular clusters, the assumption that the dynamical timescale is infinitesimal becomes questionable [Heg96]. The reason is that typical timescales of other effects fall in between the dynamical timescale and relaxation timescale.

If the orbit of a globular cluster in the galaxy is circular, the tidal field is static (except for the Coliolis force). Even in this case, the orbital timescale of the globular cluster in the galactic potential and the orbital timescale of stars around the tidal limit of the cluster are always of the same order, since the tidal radius, which is approximately equal to the Hill radius [see Equation (6.1)] with the mass of the planetesimal replaced by the mass of the globular cluster and mass of the Sun by the mass of the galaxy, implies that the average density of the globular cluster within the tidal radius and the average density of the galaxy inside the distance of the globular cluster from the center of the galaxy are the same. The dynamical timescale is determined by the mass density [see Equation (6.13)].

Thus, the effect of the rotation is dynamically important, and we cannot use the assumption of dynamical equilibrium. In the case of non-circular orbit, the dynamical effect would be even more significant, and whether we can incorporate such an effect correctly into an FP calculation has to be tested by N-body calculation.

The effect of the stellar evolution is still more problematic, since the timescale of the stellar evolution changes as stars evolve. The evolution of the stars affects the evolution of the cluster through the change of the mass of stars. Relatively light stars lose their masses through the stellar wind in the giant phase, and finally evolve into white dwarfs. More massive stars also lose some mass in the giant phase, but a much larger mass is lost in supernova explosions, which leave a neutron star behind. These are very roughly the lifecycle of a star. The gas ejected through the wind or supernova are lost from the cluster, since the potential well of the cluster is not deep enough to trap gas with a high temperature.

The lifetime of a star is primarily determined by its mass. Massive stars have a short lifetime. Thus, at any moment, some stars are losing theirs masses. Roughly speaking, the timescale of the mass loss is shorter when the cluster is young, since massive stars still exist. The lifetime of massive stars can be as short as a few million years, which is comparable to the dynamical timescale of the cluster. On the other hand, stars with a mass of 0.8 $M_\odot$, that are the most massive main-sequence stars in the present day globular clusters, have a lifetime equal to the age of universe.

Chernoff and Weinberg [CW90] set the ratio between the relaxation timescale and the stellar evolution timescale so that they mimic the value realized in real globular clusters. Since there are no adjustable timescales, this is the only possible choice as far as we use the Fokker–Planck calculation.

Fukushige and Heggie [FH95] studied the evolution of the N-body system starting from the same initial condition as used in Chernoff and Weinberg [CW90]. They, however, assumed that the evolution of the cluster in the dynamical timescale is more important than that in the relaxation timescale, and scaled the stellar evolution timescale to the dynamical timescale.

For almost all models which evaporated, Fukushige and Heggie found that the lifetime is up to 10 times longer than the result of the Fokker–Planck calculation of Chernoff and Weinberg. This rather surprising result is partly because of the better treatment of the dynamical effect in N-body simulation, but could also be affected by the difference in the relaxation timescale.

Ideally, we would like to perform N-body simulations in which both of the dynamical and relaxation timescales are correctly scaled relative to the stellar evolution timescale. Unfortunately, the only way to achieve this goal is to use the same number of stars as that in real globular clusters.

Strictly speaking, we can somewhat enlarge the relaxation timescale by using finite and large softening, but practically speaking, it is impossible to extend the relaxation timescale by a factor larger than three [Whi78], [HHM93].

It is still far too costly to directly simulate a cluster with a number of stars exceeding 10^5, even with GRAPE-4 hardware. A different approach would be to perform a series of runs with a different number of particles. If we scale the relaxation time to the stellar evolution timescale for all runs, we effectively change the dynamical timescale systematically. Thus, we might be able to extrapolate the result to a larger number of particles.

For open clusters with a much smaller number of stars, we can now perform very detailed calculations. For example, Aarseth [Aar96] described his calculation code which implements the stellar evolution for both isolated and binary stars, the effect of the finite size of the stars in close encounters, and the tidal field.

6.2.3   Algorithm and implementation

The standard individual timestep algorithm was originally developed for the simulation of star clusters. To perform the simulation of star clusters, especially after binaries are formed, we need to develop a number of special methods. In the following, we describe them in some detail.

Close encounters and compact subsystems

With the standard individual timestep algorithm, close encounters can be handled quite efficiently. However, in the simulation of star clusters, a more sophisticated treatment is necessary, for two reasons. First, the separation between two particles can be arbitrarily small, which can cause rather a serious rounding problem. Second, stable binaries pose an even harder problem.

Consider an N-body system with a half-mass radius and a total mass around unity. Here we neglect the spatial inhomogeneity. Roughly speaking, the closest distance the particles in the system experience in a unit time is $1/N$. The cross-section of the close encounter with a periastron distance of b is proportional to b, if $b << 1/N$, because of the gravitational focusing. Thus, if we integrate the system for, say, N crossing times, there would be one close encounter with a minimum separation of $1/N^2$. Note that we have to simulate the system for $O(N)$ crossing times, since the simulation must cover the thermal relaxation time.

For $N = 100\,000$, the minimum separation would be 10^{-10}. Thus, the relative rounding error is 10^{10} times larger than the normal error. Since the double precision number has only 16 digits, this would cause an unacceptably large error.

Stable binaries in star clusters have a semi-major axis of the order of $1/N$ to $1/1000N$. The periastron distance is much smaller if the eccentricity, e, follows the thermal equilibrium of $f(e) = 2e$. The orbital period of a stable binary is of the order of $1/N$ to $1/10^4 N$. Thus, we have to follow $O(N^2)$ orbits for each stable binary. Even if the computer is fast enough, the accumulation of the rounding error would become too large.

We can restore accuracy and drastically reduce the calculation cost by moving to local relative coordinates whenever necessary. Both in the case of hyperbolic close encounters and stable binaries, we replace the two particles with their center of mass and the relative motion. As a result, the relative motion of the two stars is now expressed without any loss in the effective digit. This also implies that the motion of the center of mass of the two particles is now integrated with a high accuracy.

Practically speaking, the most important advantage of this treatment is that we can skip the integration of the internal motion of the pair if they can be regarded as isolated from the rest of the system, since we can apply the analytic Kepler motion. In the case of a hyperbolic encounter, this treatment allows us to integrate the encounter with an arbitrary small periastron distance without a numerical problem. If two particles actually come close enough to each other to cause a numerical problem, we can always use an analytic solution. Of course, if three particles actually come very close, we are in trouble. However, such an event is very very rare.

In the case of binaries, this use of the analytic solution implies that we can simply update the time in multiples of the orbital period without doing anything, thereby achieving perfect accuracy and a negligible calculation cost. The only thing we need to do at the integration steps is to check if there are nearby particles, which can be done with no additional cost during the force calculation. In addition, even if the perturbation is not negligible, we can ignore most of the distant particles, since the perturbation falls off as r^{-3}. These two techniques make accurate integration of the binaries feasible, and achieve an enormous reduction in the total calculation cost [MH90].

We can also apply special techniques known as "regularization" to the integration of the relative motion of two particles. In the case of the planar motion, the regularization is expressed as follows.

The position $\mathbf{r} = (x, y)$ is transferred to a new coordinate $\mathbf{u} = (u_1, u_2)$ by

$$x = u_1^2 - u_2^2, \tag{6.25}$$

$$y = 2u_1u_2, \tag{6.26}$$

and the time t is replaced by a new variable s, defined by

$$r\,ds = dt. \tag{6.27}$$

The equation of motion

$$\frac{d^2\mathbf{r}}{dt^2} = -\frac{\mathbf{r}}{r^3}, \tag{6.28}$$

is then rewritten as

$$\frac{d^2\mathbf{u}}{ds^2} = \frac{E}{2}\mathbf{u}, \tag{6.29}$$

where E is the total energy. The Kepler problem is known to be numerically ill-behaved, in particular for large eccentricity. On the other hand, the "regularized" equation, which represents a harmonic oscillator, can be integrated with a constant stepsize, even in the case where the eccentricity e is unity.

In star cluster simulation, planar regularization is not particularly useful, since if there is any external perturbation the relative motion becomes three dimensional. In 1965, Kustaanheimo and Stiefel [KS65] invented the three-dimensional extension of the Levi–Civita transformation, which maps three-dimensional motion into four-dimensional motion. Details of the implementation are given in Aarseth [Aar85].

If a compact subsystem with more than two particles is formed, such as a binary and a single star in resonance, we cannot use the KS regularization. In 1974, Aarseth and Zare [AZ74] found a way to handle a triple system. Their basic idea is to apply the regularization to two of the three interactions, say (1-2) and (1-3) if we have particles 1, 2 and 3, so that these interactions are regularized. In this way, we can integrate the system, unless particles 2 and 3 collide. In this case, we have to switch to a new transformation.

Heggie [Heg74] generalized AZ regularization so that all three pairs are regularized simultaneously. In addition, he extended his formalism so that it can

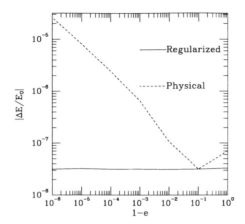

Figure 85   Comparison of direct integration and regularized integration of a binary.

handle an arbitrary number of particles. In his general N-body regularization, we have a four-dimensional equation of motion for each of the $N(N-1)/2$ pairwise potentials.

Mikkola and Aarseth [MA93b] extended AZ formalism to a general N-body system, in which only selected pairs (minimum of $N-1$) are regularized. This scheme requires the dynamical change in the regularized pair, as in the case of the original AZ three-body regularization. However, it is much faster than full Heggie formalism.

In practice, however, whether these fancy regularization techniques are really necessary or not is not clear. In the case of the binary system, the regularization does save calculation cost, in particular if the eccentricity is large. Figures 85 and 86 show the error in the total energy and the number of timesteps for the Kepler problem. The integration scheme is fifth-order Runge–Kutta–Fehlberg with automatic stepsize control. The accuracy criterion is that the absolute error of all variables is smaller than 10^{-10}. In the case of regularized integration, we can see that the energy is quite well conserved and that the number of timesteps does not depend upon the eccentricity. On the other hand, in the case of the direct integration the calculation cost increases quite significantly, and the error also becomes larger.

In direct integration of highly eccentric binaries, practically all the integration error is generated around the pericenter, where the velocity is the largest. For eccentricity of less than, say, 0.99, the degradation in accuracy is quite modest, and could be recovered with a slight increase in the calculation cost. The increase in the calculation cost is also not very large, in particular if

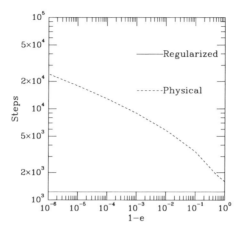

Figure 86   Same as Figure 85, but plotting the number of timesteps per orbit.

we take into account the additional overhead associated with the coordinate transformation.

In the case of compact systems with more than two particles, the use of general N-body regularization alone is not sufficient, since in many situations hierarchical systems are formed. In fact, the only stable configuration of few body systems are hierarchical configurations, where at each level the system can be regarded as a weakly perturbed binary system. For such hierarchical configurations, the natural way is to map the subsystem to a binary tree.

In the binary tree, the motion of a node is expressed as the motion relative to its parent node. If a tree has more than two levels, the absolute coordinates of a node can be calculated by tracing the tree upwards to the top level. The change of this relative coordinate is integrated directly, using the Hermite scheme. Since the positions of a node and its counterpart in the tree are redundant, we actually integrate one of them and convert it to the coordinate of the other.

How does one calculate the force on the low-level node? By definition, the position of a low level node is relative to its parent. Therefore, the force must be relative to its parent as well. To be more specific, the acceleration of particle k in global coordinates is expressed as

$$\mathbf{a}_{g,k} = \sum_{j \neq k} \mathbf{a}_{jk}, \qquad (6.30)$$

where $\mathbf{a}_{jk}$ is the acceleration of particle k due to the gravitational field of particle j. If we consider the case of a node i, whose two children are k and l, the acceleration of node i is expressed and calculated as

$$
\begin{aligned}
m_i \mathbf{a}_i &= m_k \mathbf{a}_k + m_l \mathbf{a}_l \\
&= m_k \sum_{j \neq k} \mathbf{a}_{jk} + m_l \sum_{j \neq l} \mathbf{a}_{jl} \\
&= \sum_{j \neq k,l} m_k \mathbf{a}_{jk} + m_l \mathbf{a}_{jl}.
\end{aligned}
\tag{6.31}
$$

The acceleration of k can be rewritten as

$$
\mathbf{a}_k = \mathbf{a}_{lk} + \sum_{j \neq k,l} \mathbf{a}_{jk}.
\tag{6.32}
$$

Thus, the acceleration of a node relative to its parent is calculated as

$$
\begin{aligned}
\mathbf{a}_k - \mathbf{a}_i &= \mathbf{a}_{lk} + \sum_{j \neq k,l} \mathbf{a}_{jk} - \frac{1}{m_i} \left(\sum_{j \neq k,l} m_k \mathbf{a}_{jk} + m_l \mathbf{a}_{jl} \right) \\
&= \mathbf{a}_{lk} + \frac{m_l}{m_i} \sum_{j \neq k,l} \mathbf{a}_{jk} - \mathbf{a}_{jl}.
\end{aligned}
\tag{6.33}
$$

Until now, we have assumed that k and l are simple particles. However, actually the above reasoning is valid no matter whether they are particles or composite nodes.

To put it in a slightly different way, the force on a composite node in the absolute coordinate is simply the sum of the forces from all particles which do not belong to that node to all particles under that node. The relative motion of a binary component is the force from its counterpart plus external perturbation.

Even if the compact system is not stable, we can still force a binary tree structure without a significant loss in accuracy. The possible loss in accuracy comes from the frequent change of the structure of the tree. This can be suppressed by introducing some hysteresis factor. To be more specific, the rules for the change of the binary tree structure are summarized as

(a) A new tree is formed if two particles (or the center of mass of an already formed tree) are closer than a prescribed distance r_{close}.
(b) The top level tree is disrupted if the separation of the two components is larger than γr_{close}, where γ is a parameter larger than unity.
(c) For a particle in a tree, if the distance to its nearest neighbor is less than γ times the distance to its counterpart in the tree structure, the tree structure is changed so that it forms the tree with its nearest neighbor.

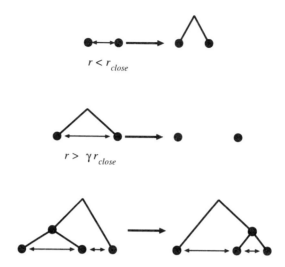

Figure 87   Three rules to change the tree structure.

We have developed a time integration program for a globular cluster which makes extensive use of binary tree structure to express compact subsystems, and always uses direct integration instead of regularization techniques. The main advantage of this scheme is it's simplicity. The new program can handle arbitrary structures of an arbitrary number of particles, either hierarchical or democratic. On the other hand, with the approach of using general N-body regularization, we need additional code to handle hierarchical systems. In Aarseth's NBODYx programs, a special technique for hierarchical triples was implemented. However, it is rather difficult to extend it to hierarchical quadruples or more complex systems. The approach based on the binary tree structure can naturally handle arbitrary complex systems. Figure 88 shows a tree of eight particles formed during the simulation of a 10-body system. Each filled point denotes a particle and an open circle a composite node. This tree is created with too large an r_{close}, to demonstrate that a complex tree can be formed and time integration works for a tree like this.

The Neighbor scheme

In the individual timestep algorithm, each particle in the system has its own timestep. We can extend this idea to all $N(N-1)/2$ pairs, and think of a scheme in which each pairwise interaction has its own timestep.

In practice, such a scheme would be too complicated to implement. Ahmad and Cohen [AC73] developed a more practical scheme, in which the force

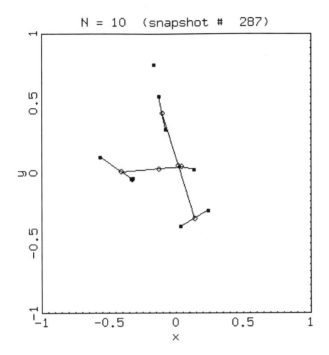

Figure 88   The tree structure constructed dynamically.

on a particle is divided into two components, one from its neighbors and the other from distant particles. These two components are updated using different timesteps, Δt_n for the neighbors and Δt_d for distant particles. Thus, at certain timesteps we calculate only the forces from neighbors (neighbor step), while at other timesteps we calculate both the forces from neighbors and distant particles (distant step). Whether a particle is a neighbor or not is determined by the distance. If this particle is within the distance r_n from the particle for which we calculate the acceleration, it is included in the list of neighbors.

Detailed descriptions of the algorithm and analysis of the calculation cost are given in Makino and Hut [MH88] and Makino and Aarseth [MA92]. To summarize their findings, the relative advantage of this algorithm over the original individual timestep algorithm is $O(N^{1/4})$, with the coefficient of order unity. However, to actually achieve this optimum result, the memory needed to store the neighbor list is $O(N^{7/4})$. In general, if the amount of memory available is $O(N^{1+\alpha})$ with $\alpha < 3/4$, the gain in the calculation speed is $O(N^{\alpha/3})$.

Thus, on a general-purpose sequential computer, the gain one can achieve is fairly large. For $N = 10^4$, we can get a speedup of around a factor of 10.

The actual gain one can achieve on present-day high-performance computers, in particular distributed-memory parallel computers or GRAPE type machines, is not so large. The main reason is simply that the reduction in the calculation cost also implies a reduction in the degree of parallelism. Thus, the efficiency of calculation of the neighbor force is not as high as that of the total force. On vector machines, the fact that the length of the neighbor list is relatively short implies that the startup time of the vector operation becomes quite significant, resulting in a large loss in the efficiency. Moreover, the memory access during the force calculation becomes irregular, since we only access the particles in the neighbor list. On vector machines, irregular memory access tends to be much slower than regular memory access. On cache-based machines, this loss is actually even more significant, since the irregular access implies that a large fraction of the cache line read in from the main memory is simply discarded.

In the case of GRAPE, we have not yet found an architecture which can run the Ahmad–Cohen scheme efficiently, and at the same time achieve massive parallelism and relatively low memory requirement.

The McMillan–Aarseth individual-timestep treecode

McMillan and Aarseth [MA93a] developed a high-accuracy treecode with an individual timestep. A similar attempt was also made by Jernigan and Porter [Jer85] [JP89], but the main difference between the two approaches is that Jernigan and Porter used the binary tree to represent the whole system, in the same way as we described earlier. On the other hand, McMillan and Aarseth used the Barnes–Hut octree. As a result, Jernigan and Porter expressed the nodes in the coordinates relative to their parents, and integrated the motion of all nodes in the tree as well as that of real particles, while McMillan and Aarseth used the tree to calculate only the force. They used the usual KS-regularization to handle close encounters and stable binaries.

Here, we make a brief comparison of the tree algorithm on a general-purpose parallel system and direct summation on GRAPE. If we use a single-processor machine, it is clear that the tree algorithm is more efficient for a sufficiently large value of N. McMillan and Aarseth found the cross-over point to be around $N = 1000$. They also found that the scaling of the calculation cost of the direct summation code is $\sim N^{2.16}$, while that of the treecode is $\sim N^{1.6}$. Thus, the relative speedup over the direct summation would be around $N^{0.5\sim0.6}$, for N not too much larger than 1000. If we can extrapolate this scaling to $N = 10^6$, the gain in the calculation cost would be a factor of $30 \sim 50$. This gain is not considerably larger than the gain one can achieve with the Ahmad–Cohen scheme. Taking into account the relative complexity of the treecode over direct summation or the Ahmad–Cohen scheme, it looks unlikely that treecode will be the method of choice for simulations of globular clusters.

In theory, one could achieve further speedup by combining the treecode with the Ahmad–Cohen algorithm. This combination, in theory, could give another factor of $5 \sim 10$ speedup for $N = 10^6$. However, even with this speedup, we would still need multi-teraflops sustained performance to finish one simulation of a 10^6-body system in less than a year. Given the complexity of the algorithm and inherently low level of parallelism, the chance that the treecode with an individual timestep can achieve more than a small percentage of the theoretical peak performance of a massively parallel computer is rather low.

6.3 Galactic nuclei

The study of the galactic nucleus is too big a field to be covered in a section of this small book. We limit our attention here to a rather small problem, namely the stellar dynamics of the system with massive central black hole(s). To really understand the galactic nucleus, the inclusion of the gas dynamics coupled with radiative transfer, star formation and stellar evolution would be crucial. However, before proceeding to a study of such complex systems, we first need to have a clear understanding of the stellar dynamics. Of course, there are many studies which include gas dynamics and other physical processes. Just to show one example of work performed using GRAPE hardware, Taniguchi and Wada [TW96] studied the dynamics of a black-hole binary-gas disk-stellar bulge system using SPH. They found that the presence of a black-hole binary can cause strong spiral patterns in the gas disk, which would cause the gas to fall to the center, triggering the nuclear starburst. In the following, however, we mainly discuss the study of pure stellar dynamics that we have been studying.

6.3.1 Black hole binaries

The possibility of the formation of massive black hole binaries at the cores of elliptical galaxies was first pointed out by Begelman et al. ([BBR80], hereafter referred to as BBR). Their argument is summarized as follows. The most likely energy source for AGN (Active Galactic Nucleus) or QSO (Quasi-Stellar Object) activities is massive black holes with the mass $M_{BH} \simeq 10^8 M_\odot$. If two galaxies with central Massive Black Holes (MBHs) merge with each other, the MBHs will sink to the center of the merger remnant, because of the dynamical friction from stars, and form a binary.

BBR argued that such a binary would have a typical lifetime larger than the Hubble time, since the BH binary would create a "loss cone" in the distribution of field particles, which is repopulated only in the thermal relaxation timescale of the core.

Ebisuzaki *et al.* [EMO91] pointed out that the lifetime is much shorter if the BH binary is highly eccentric. BBR assumed that the binary becomes circular. The timescale of the gravitational radiation is proportional to $(1 - e)^{-3.5}$ for $e \sim 1$, where e is the eccentricity. If the BH binary becomes highly eccentric, therefore, the lifetime would become much shorter. In fact, even for the average "thermal" eccentricity of 0.7, the lifetime is shorter than that of a circular binary by nearly two orders of magnitudes. Fukushige *et al.* [FEM92] performed a simulation of the evolution of the BH binary under the assumption that the interaction between a BH and field stars is described by the standard dynamical friction theory [Cha43]. They found that eccentricity of a BH binary increases as the binary evolves.

Of course, it is not likely that the standard theory of dynamical friction can be applied to the evolution of the binary without any change. The standard theory assumes that the massive object orbits in free space, and the effect of each field star is described by means of an independent hyperbolic 2-body encounter in free space [Cha43]. Each component of the binary orbit around each other, which means that the encounters with field stars cannot be regarded as encounters in free space, except for the limiting case where the incoming velocity of the field star is much larger than the orbital velocity of binaries. More detailed study was clearly necessary. Mikkola and Valtonen [MV92] performed a Monte-Carlo experiment to determine the differential cross-section for the change of the energy and eccentricity of the BH binary. They concluded that there would be a modest increase in the eccentricity, but not as large as that predicted by Fukushige *et al.* [FEM92]. Quinlan [Qui96] performed a much more detailed analysis and obtained a similar result.

Makino *et al.* [MFOE93] performed a self-consistent N-body simulation of the evolution of a BH binary in the core of a galaxy with 16 384 particles. This calculation was performed using GRAPE-3, with a second-order individual-timestep algorithm, described in Section 5.1.4. As far as the eccentricity is concerned, their results were consistent with that of the Monte-Carlo studies. If the initial eccentricity is large, it can go up. However, if the initial eccentricity is small, it remains small.

However, they obtained rather an intriguing result: the evolution of the separation of the binary was not halted even after the separation became less than 1/10 of the critical radius obtained by BBR.

Makino [Mak97] repeated the calculation of Makino *et al.* [MFOE93] with GRAPE-4, using up to 256k particles. Figure 89 shows the time evolution of the binding energy of the BH binary for runs with a different number of particles. Here, solid, short-dashed and long-dashed curves represent results

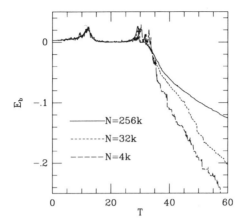

Figure 89   The time evolution of the binding energy of binaries in simulations
of merging two galaxies with central black holes.

with 256k, 32k and 4k stars per galaxy. Initial conditions are the same except
for the number of particles. The main result is that the growth rate of the
binary does depend upon the number of particles. Thus, the loss-cone depletion predicted by BBR actually takes place. However, the dependence of the
growth rate to the number of particles is entirely different from the prediction
by BBR.

According to BBR's theory, the loss cone is refilled through two-body relaxation between stars outside the loss cone. Thus, the timescale of the refilling
is proportional to the two-body relaxation time, and therefore the growth
timescale of the BH binary would also be proportional to the number of particles N. The numerical result shows a much weaker dependence, which seems
to indicate that the growth timescale is proportional to $N^{0.3}$ [Mak97].

The numerical result seems solid, but the reason for the discrepancy between
the theory and the numerical result is not well understood. Rather, there are
too many possible explanations, all of which sound plausible. In the following,
we list just a few of them.

The very assumption that the loss cone is formed might be wrong, if the
average change of the angular momentum per orbital period of a typical star
is larger than the maximum angular momentum of the orbit in the loss cone
[LS77]. In Fokker–Planck approximation, the change in the angular momentum per orbital period is infinitesimal. Thus, the loss cone is clearly defined
as the absorbing boundary condition in the (E, J) plane. In reality, however,
the change is finite, and can be larger than the loss-cone radius. If the change
in the angular momentum per orbit is large, one particle can go into and out

of the loss cone in a single orbit. In this case, the loss cone is refilled in a timescale shorter than the depletion, and we see only a modest decrease, if at all visible, of the central density.

The assumption that the BH binary lies at the exact center of the system might also be wrong. In our simulations, the mass ratio between the field stars and BHs is at the maximum 3×10^3. Thus, the average random velocity of the BH binary is around $1/\sqrt{3 \times 10^3} \simeq 2\%$ of that of a typical star, and the BH binary wanders around the exact center. This increases the effective radius of the loss cone as well as the depletion timescale.

The relaxation timescale of the field stars around the BH binary might be significantly shorter than the standard relaxation time, at least for the angular momentum [RT96]. If the potential field is almost Keplerian, the orbits of stars are also nearly Keplerian. In other words, the orbits of stars are approximately elliptic and closed. This fact implies that the resonance between orbits of a similar orbital period enhances the relaxation. In a normal stellar system, the orbits of stars are not closed, and resonance is of little significance. Rauch and Tremaine [RT96] called this effect "resonant relaxation". This effect could reduce the timescale of the refilling quite drastically.

Finally, the energy exchange between the stars in the orbit close to the loss cone and the random motion of the BH binary could reduce the timescale of the refilling further.

Practically speaking, the dependence of the evolution timescale found by Makino [Mak97] is strong enough that the evolution of the BH binary in a real elliptical galaxy would be effectively halted. Thus, the probability that the massive central black hole is actually a binary system might be fairly high. Thus, we are led to a prediction that the majority of massive central black holes might be binaries. One theoretical problem with such a picture is that if two galaxies with central BH binaries, or one with a binary and the other with a single BH, merge, the binary-single BH interaction would most likely eject the single BH, and in some cases the binary as well, from the merger galaxy.

This possibility of the ejection of the central BHs through binary-single BH interaction has been studied in an entirely different context. Lacey and Ostriker [LO85] proposed massive black holes of a mass of around $10^6 M_\odot$ as the constituent of the dark matter in the galactic halo. This theory has the nice implication that it explains well the observed random velocities of stars in the galactic disk. It is well known that older stars have larger random velocities (here, random velocity accounts for the deviation from the circular orbit, in the same way as in the planetesimal dynamics). The thermal relaxation caused by individual stars is too small to account for the large velocity dispersion of old stars. Thus, we need some massive objects to "heat up" stars.

Another old problem of the galactic disk is that there seems to be a huge amount of mass somewhere around the disk. The observation of the rotation velocity of stars in our galaxy, as well as that of many other spiral galaxies, has made it clear that there is significantly more mass in the galaxy than can be observed in the form of stars and interstellar gas. In addition, numerical simulation of disks has shown that self-gravitating disks are unstable against bar instability [OP73]. Thus, the standard theory of the structure of the galaxy at present is that there is "dark matter" of a mass comparable or larger than the mass of the disk distributed more or less spherically. This distribution of mass is called the "dark halo".

What makes up this dark halo is one of the central problems of modern astrophysics, and researchers have tried every explanation, starting from neutrinos with a finite mass, exotic elementary particles which interact with everything else but only very weakly, Jupiter-sized planet-like objects floating around, brown dwarfs, and black holes with various masses, which could have been formed in the early universe. The postulate by Lacey and Ostriker [LO85] is about the most exotic, but it can explain several important observational facts.

If the galactic halo is full of massive black holes, as suggested by Lacy and Ostriker, some of them would fall into the galactic center. They estimated that the mass which could have fallen into the core within the age of the universe is around $10^8 M_\odot$. This estimate contradicts the current observational upper limit of the mass of the central black hole in the galaxy, which is only $3 \times 10^6 M_\odot$. To circumvent this difficulty, Lacy and Ostriker came up with the idea that all of these black holes, except for the possible one remaining binary, have been ejected out through binary-single BH interaction. Xu and Ostriker [XO94] even performed an N-body simulation of these massive black holes, under the fixed potential well of the galactic disk and bulge stars and additional dynamical friction.

Whether such an ejection actually occurs in our galaxy or in merger remnants is not clear, though. Hut and Rees [HR92] pointed out that an ejection with 100% efficiency is quite unlikely for the case of our galaxy, simply because of the statistical fluctuations. They also argued that once a massive BH is formed through the merging of two BHs due to gravitational wave radiation, the ejection of that massive BH becomes much harder.

Makino and Ebisuzaki [ME94] made a simple analytical estimate of the probability of merging, taking into account the distribution of the eccentricity of the BH binary. They assumed that the eccentricity follows the "thermal" distribution achieved through a number of weak encounters with field stars [Heg75], which might grossly underestimate the eccentricity of the binary formed by strong three-body scattering. Thus, their result must be regarded as the lower bound of the probability of merging. Even so, they could show that the merging probability is non-negligible in the case of BHs in our galaxy, and practically equal to one for massive BHs in ellipticals.

Their result implies that the ellipticals can contain such binaries, but that if such a binary interacts with another BH, two out of three BHs are likely to merge through gravitational wave radiation, resulting in a new BH binary.

6.3.2   Structure of galactic nuclei

Present elliptical galaxies do not contain much gas. It seems natural to assume that they are formed through a dissipationless process, or through some dissipational process which consumed most of the gas.

In the standard cosmology, anything with mass larger than the critical mass is formed by the merging of the smaller mass, and as the mass becomes larger, the formation epoch becomes later. So it is reasonable to assume that giant ellipticals are formed through the merging of smaller galaxies which do not contain a very large amount of gas. In fact, the merging hypothesis, first proposed by Toomre [Too77], has been quite successful in explaining many of the observed properties of large elliptical galaxies [BH92], [OEM91].

However, the merging hypothesis has one serious problem: The structure of the central region predicted by the merging is completely different from the observed structure.

The problem is the following. There seems to be a strong correlation between the effective radius and the core radius of large elliptical galaxies, which, very roughly speaking, means that all large ellipticals have one universal structure (see, for example, Lauer [Lau85]). However, this observation contradicts both the theory and numerical results, if we assume that large ellipticals are formed by the merging of smaller galaxies. The theory suggests that the central phase space density, the value of the distribution function f at the center and $v = 0$, is kept roughly constant through merging [Car86]. If the central velocity dispersion is not changed by merging, therefore, the core radius should be roughly the same. Simulations of the merging of two galaxies have shown that this theory is correct. In merging simulations, the core radius is always kept unchanged by merging [FSD83], [OEM91].

To make matters even more mysterious, high-resolution ground-based observations have revealed that the cores of large ellipticals cannot be modeled by an isothermal distribution. Finally, recent HST observations have demonstrated that "cores" are in fact all cusps with the density profile $\rho \sim 1/r$ or shallower [LAB+95], [FVF+94], [GRA+96].

All merging simulations which demonstrated that the core size is unchanged are either pure N-body or with gas dynamics, but did not included the effect of the central BH. Since a large fraction of ellipticals might have central BHs whose mass is comparable to that of the "core", its effect can dominate the dynamics of the structure of the central region. In fact, it turned out to be difficult to model the shallow cusps as suggested by recent HST observations, without assuming a large central mass.

The available theories for the structure and evolution of galaxies with central black holes, however, failed to explain the HST observation. Young [You80] considered the scenario where the BH mass grows slowly in the center of a spherical galaxy with a flat core. He showed that the density profile approaches $\rho \propto r^{-1.5}$. Sigurdsson *et al.* [SHQ95], through N-body simulation

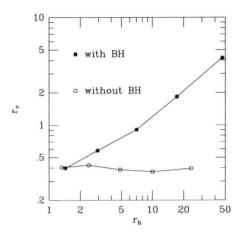

Figure 90   The relation between the core radius and the half mass radius, for growth of a galaxy through merging.

using a variation of the Hernquist–Ostriker $O(N)$ method [HO92], obtained essentially the same result.

Ebisuzaki *et al.* [EMO91] proposed the idea that energy generation from the binary BH heats up the particles around, resulting in an increase in the core radius. This hypothesis explains the observed correlation between the core radius and the effective radius quite naturally. The core radius relative to the effective radius would be determined by the ratio of the BH mass and the total mass of the galaxy. If large ellipticals are formed through merging, it is natural to assume that the mass of the central BH and the total mass of the galaxy are correlated. Therefore, larger ellipticals have larger cores.

Ebisuzaki *et al.* [EMO91] performed merging simulations with up to 4096+2 particles using GRAPE-2. They demonstrated that the core radius actually becomes larger. Makino and Ebisuzaki [ME96] performed the simulation of hierarchical merging, in which the product of one merger is used as the progenitor of the next merging simulation. They used GRAPE-4 and 32 768 particles. They showed that the ratio between the core radius and the effective radius converges to a value which depends upon the mass of BHs. Figure 90 shows the relation between the core radius and the half-mass radius (which is close to the effective radius) for repeated merging simulations. The sequence of mergings of galaxies with central black holes shows that the core radius is roughly proportional to the half mass radius, while in the merging without central black holes, the core radius remains almost constant.

Moreover, the density profile of the merger product obtained by Makino and Ebisuzaki [ME96] indicates that the central region is not a flat core but a

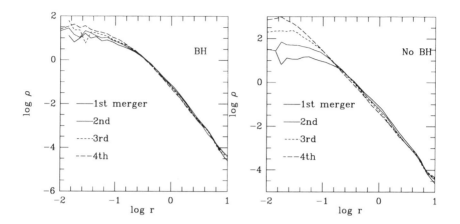

Figure 91   The normalized density profiles of merger remnants. (Reproduced
                from Makino and Ebisuzaki (1996).)

cusp with $\rho \sim 1/r$. Thus, the merging of galaxies with a central BH explains
the observed structure of ellipticals quite well. Figure 91 shows the evolution
of the density profile for merging runs with and without black holes.

However, it turned out that 32k particles were not enough. Makino [Mak97]
repeated essentially the same calculation as performed by Makino and
Ebisuzaki [ME96] with 256k particles, and found that the structure of the
core is different from that obtained in 32k runs (see Figure 92). The most important difference is that the volume density actually goes down in the vicinity
of the BH binary in the 256k runs. The "loss cone" is actually observed.

6.3.3   Algorithm and implementation

The direct method

The calculations performed on GRAPE-4 all used a simple direct method
with an individual timestep, with only one modification. The gravitational
force from the BH particles is calculated on the host computer. This is mainly
because we want to use pure $1/r$ potential for the force from BH, while we
want to use a finite softening for interactions between field particles.

When the BH binary becomes very tight, special techniques, as described
in Section 6.2.3, might be of some help. However, in most calculations we have
performed so far, this would not result in a large reduction in computer time,
since the orbital period of the BH binary is relatively long.

The mass of the BH particles is around 1% of the total mass of the system
in our simulations. Thus, if the semi-major axis is 10^{-3}, the binding energy of

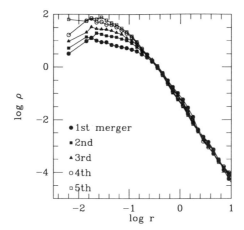

Figure 92   The normalized density profiles of merger remnants of simulations
with 256k particles.

the binary is around 10% of the total energy of the system. In this case, the
timestep for the BH particle is around 10^3 times smaller than that of typical
particles in the system. Therefore, the calculation cost of the BH particle is
not dominant if $N > 10^3$.

Hybrid methods

Even with GRAPE-4, we can handle only up to a quarter of a million particles
if we want to finish one simulation in a reasonably short time (less than a
month). On the other hand, to obtain a reliable result for the evolution of the
BH binary and the structure of the core, we need more than 10^7 particles,
which implies a Petaflops machine if we use the direct method.

In the case of the star cluster, which would need a Petaflops machine for
10^6 particles, the use of schemes other than the direct method is not really
practical. This is partly because the required accuracy is high, and partly
because the distribution of the timestep is quite ill-behaved. In the case of the
simulation of the galaxy, however, the situation is different.

The duration of the simulation is short. Typical galaxies have an age of
around 100 dynamical times, compared to more than 10^4 dynamical times of
typical globular clusters. In addition, we can use the potential softening. In
fact, we should use the softening to reduce the relaxation effect. Thus, schemes
like the treecode and other fast algorithms have a better chance.

It should be noted that we still cannot use the standard treecode with a
constant timestep, because the timescales of BH particles and field particles

close to BH particles are very short. Thus, some kind of hybrid approach might be most useful. Quinlan [Qui96] has been working on such a hybrid between the direct method and the SCF method for quite some time. Another possibility is to use the tree with an individual timestep and high-order time integration [MA93a]. To combine the tree algorithm, the individual timestep and the GRAPE hardware is, however, not easy. As discussed in Section 6.2.3, the implementation of treecode with an individual timestep on a general-purpose parallel computer is already a complex problem, essentially because of the low degree of parallelism.

In the following, we give a rough idea of possible schemes in the case of a general-purpose computer and GRAPE hardwares. None of them has been implemented so far.

A combination of the direct and tree algorithms

The goal here is to reduce the total calculation cost by making use of some form of the tree algorithm without completely sacrificing the simplicity of the direct summation algorithm, which makes it possible to use GRAPE to obtain a high speed.

There can be many different ways to combine the two algorithms, depending on how we divide the work. One way is to divide particles by the timestep requirement. If the timestep of a particle is equal to or longer than a critical value, the force from it is calculated using the tree, and if the timestep is smaller, the force from it is calculated directly. In this case, we have to change the tree structure only at the interval of this critical timestep.

If we can choose this critical timestep in such a way that it is not much smaller than the average timestep of particles in the system, and yet there are not too many particles with the timestep less than this critical timestep, we could achieve a significant speedup over the direct summation.

One problem is how we calculate the force on the particles with a small timestep from the tree. To make use of the tree, GRAPE should perform some indirect addressing, or the host computer must handle this part. Which would be the best strategy depends upon the size of the problem, the distribution of the particles, and the ratio of the speed of the host computer and that of the GRAPE hardware.

A combination of direct and spherical harmonic expansion type algorithms

If we are interested in the evolution of the BH binary in a single galaxy, we might be able to use some sort of expansion scheme (for example, Hernquist and Ostriker [HO92]) to handle the outer part of the system. Here again, the question is how to divide the work between two schemes. The simplest would be to have a critical radius, but in this case the switching between the two schemes might cause some undesirable effects like systematic drift of the total energy. Another possibility is to allow two schemes to overlap, as in the case of the P^3M scheme. Yet another possibility would be just to use the timestep criterion.

6.4   Galaxy interactions

6.4.1   Encounters between isolated galaxies

Interacting galaxies have been one of the foci of research into galactic stellar dynamics. For encounters and mergings of two galaxies, there have been a number of excellent works by Barnes, Hernquist and their collaborators [BH92].

Studies on the interaction of galaxies using GRAPE have so far concentrated on simpler cases, such as the interaction and merging of two spherical galaxies. The Marseille group has started to study the interactions between spiral galaxies and between a spiral and its small companion [APB97b], and more research will follow in this direction. However, in this book we concentrate on earlier work.

Okumura *et al.* [OEM91] investigated the structure of the remnants of merging two Plummer models from parabolic orbits. Calculation was performed on GRAPE-1, and the total number of particles was 16 384. Their main conclusion is that the ellipticity and rotation velocity of the merger remnants agree well with the observation. They obtained the non-dimensional rotation velocity defined as V_{max}/σ_0, where V_{max} is the peak of the rotation velocity and σ_0 is the velocity dispersion at the center. This value depends strongly upon the initial periastron distance, but saturates at ~ 0.6 for mergings from a very large initial periastron separation.

Their result is in good agreement with the observation of large ellipticals, which shows rather a sharp cutoff in the distribution of V_{max}/σ_0 at around 0.6. On the other hand, simplified theoretical argument [Whi79] suggested that many merger remnants should rotate more rapidly. The basic assumption of White [Whi79] is that the merger remnant retains most of the orbital angular momentum of the initial condition. This assumption is, however, not quite true, as demonstrated by Sugimoto and Makino [SM89], who studied

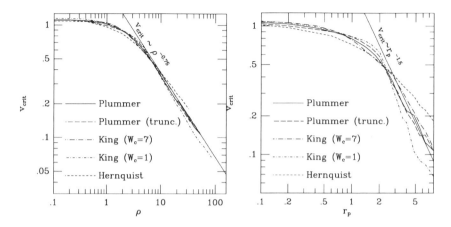

Figure 93 Critical velocity for merging as a function of (left) impact parameter at the infinity, and (right) periastron separation.

the merging process of binary globular clusters. Most of the orbital angular momentum is carried away by stars which escape from the system. Though the total number of escaping particles is small, they can carry away a large amount of angular momentum because they are very efficiently accelerated by the rotating potential field.

Makino and Hut [MH97] studied the merging cross-section of spherical galaxies. Though the merging is of critical importance in understanding the evolution of galaxies in clusters, no reliable result on the merging cross-section has been available. In the late 1970s, there were several studies to determine the merging cross-section of two galaxies, both by direct N-body simulation and by an analytic approach. However, N-body simulation then could handle only ~ 100 particles, in particular if a large number of runs was necessary.

Makino and Hut [MH97] used 2048 particles per galaxy for most runs, and up to 16 384 particles in some cases. They used a GRAPE-3A board and a simple direct summation algorithm.

Figure 93 shows their result, in terms of the critical relative velocity at an infinite distance for merging as a function of the impact parameter at infinity and the periastron distance. They performed the simulation for several different galaxy models. In figure 93(a), the difference between models seems to be rather small, while in figure 93(b) there is rather large difference. The reason why the apparent difference is small in figure 93(a) is simply that the small change in velocity at the infinity, for the same impact parameter, can cause a large change in the periastron distance.

Practically all theoretical estimates on the merging cross-section relied on the so-called tidal impulse approximation, where one calculates the energy exchange between the orbital motion and internal motion assuming that the force is expressed by the first order tidal force, and yet assuming that one can apply the impulse approximation.

Both assumptions are questionable. The use of the first order tidal force is valid if the galaxies have finite radii sufficiently small compared to the periastron distance. Many realistic galaxy models, however, have power-law halos, such as the r^{-5} halo in Plummer model and the r^{-4} halo in the Hernquist or Jaffe models. The use of impulse approximation is certainly invalid, since the timescale of the encounter is *much longer* than the typical orbital timescale of the stars in the galaxy, for the critical velocity. However, studies in late 1970s and 1980s demonstrated a reasonable agreement between the result of the numerical simulation and that of tidal impulse approximation. We have shown that this agreement is not real. In fact, we found much better agreement with an approximation which assumes only particles around the radius comparable to the periastron separation are affected by the encounter.

In a typical cluster of galaxies, the typical relative velocity between galaxies is several times larger than the internal velocity dispersion of the galaxies. In this environment, merging of two galaxies only occurs when two galaxies collide with relative velocity significantly lower than the typical relative velocity. This implies that we can use the low-velocity end of the velocity distribution function to estimate the merging rate of galaxies in clusters. Makino and Hut [MH97] determined the merging rate under this assumption.

Funato and Makino [FM97b] performed a systematic survey of the effect of high-speed, non-merging encounters on the structure of galaxies in the cluster.

It has been long known that the structural parameters of elliptical galaxies in rich clusters show marked correlation. The best known is the existence of the fundamental plane, whose essential meaning is that the mass-luminosity ratio (M/L) of elliptical galaxies show very weak dependence on other parameters, such as the total luminosity [DD87].

The Faber–Jackson relation and the $D_n - \sigma$ relation indicate a further correlation between structural parameters. Both imply there is positive correlation between the total luminosity (or, equivalently, the total mass within the effective radius) and the effective radius itself. In the case of the Faber-Jackson relation, it is expressed as

$$L \propto \sigma^4, \tag{6.34}$$

where L is the total luminosity of an elliptical galaxy and σ is its velocity dispersion.

How such a correlation developed is an important question, for the following reasons. First, it might give some clue to the formation process of the ellipticals. If the relation is primordial, it gives a strong constraint on the formation process. Secondly, relations of this type have been used to determine the age of the universe. The age and the fate of the whole universe are certainly the single most important problem in modern astronomy.

The key to determining the age of the universe is the Hubble constant, which measures how rapidly the whole universe is expanding. Hubble first suggested that galaxies are receding from us and that receding velocity is apparently proportional to the distance to galaxies. The most natural interpretation of this finding is that the whole universe is uniformly expanding. In fact, the general relativity theory has such an expanding solution.

It is not impossible to construct a non-expanding, stationary universe with general relativity, but it requires the non-zero cosmological constant, which, roughly speaking, "supports" the whole space so that it does not collapse onto itself. It does not sound too plausible to assume that the value of the cosmological constant is just enough to keep the universe from collapsing or expanding.

The expansion implies that the universe was smaller and more dense in the past, and that the universe was "born" at a certain time. In other words, the universe has a finite age.

The Hubble constant, which is the receding velocity divided by the distance, directly determines the age of the universe, though there still remain some uncertainties caused by the average mass density.

To obtain a reliable value for the Hubble constant is the central goal of observational cosmology. To measure the Hubble constant, we need to determine the distance and receding velocities of distant galaxies. We want to use galaxies as far away as possible, since the velocities of nearby galaxies are affected by the local inhomogeneities of the universe, such as clusters of galaxies or the large scale structure.

To determine the accurate value of the receding velocity, or, in the more usual term, the redshift, is relatively easy, since it can be determined directly from the measurement of the locations of absorption lines in the spectra (hence the name "redshift").

To determine the distance is much harder, since there is no simple direct way of measuring it. We can determine the distance to the stars in the solar neighborhood through a parallax measurement. Recently, accurate distance data from the HIPPARCOS satellite [Ber85] has extended the limit quite significantly. Even so, there is no hope of measuring the distance of anything outside our galaxy using parallax. Any measurement of distance to extragalactic objects relies on some sort of "standard candle", or the hierarchy of the candles.

The basic idea of the standard candle is quite simple. If we know that astronomical objects of a certain type have a unique absolute luminosity, we can determine the distance from their observed luminosity. For nearby galaxies, we can use the luminosity of a special type of variable star as the candle, since their luminosities are known to be similar. However, for very distant galaxies, we cannot resolve single stars, and determination of the distance is more difficult.

If there is a relation between the luminosity and some other observable quantities of galaxies, we can use it as the standard candle. The Faber-Jackson relation is one example. If the relation is universal among galaxies in different clusters, it can serve as the distance measure.

Whether the FJ relation is a universal relation or not is difficult to determine. Within each cluster, the existence of the relation is fairly clear. It is difficult to see if ellipticals from different clusters are on the same line or not, simply because we do not know the distance to each cluster. Thus, numerical studies of the evolution of galaxies in the clusters could play some role.

Funato *et al.* [FME93] performed direct simulation of the clusters in which galaxies are expressed as an N-body system. We will provide more detail in Section 6.4.3. Here, it is sufficient to say that they found that the FJ relation develops naturally as the result of the evolution of galaxies in the cluster, driven by the interactions with other galaxies and the effect of the tidal field of the parent cluster.

Funato and Makino [FM97b] performed a detailed study of the change in the velocity dispersion and the mass of individual galaxies through encounters. In the late 1970s until the mid-1980s, there were several studies, either using N-body simulation or semi-analytic approaches, to determine the effect of the encounters on the structure of the galaxies.

N-body simulations in the early 1980s could use only a small number of particles. Small N implies that the galaxies relaxed significantly during the encounter. Moreover, it was a usual exercise to apply the "annealing" to the model galaxy before using it, to guarantee that it is already thermally relaxed.

This annealing did reduce the change of the structure of the galaxy during encounter experiments, but quite naturally changed the initial structure of the galaxies. Real elliptical galaxies are not thermally relaxed. As will be discussed in the next section, the luminosity profile of typical ellipticals are well approximated by Hernquist or Jaffe models which have r^{-4} halos, while annealed model galaxies had halos much closer to r^{-3}. This halo developed as a result of the thermal relaxation around the half mass radius of the model galaxies. The high-velocity tail of the distribution function develops through the relaxation, which forms the halo.

Funato and Makino [FM97b] solved this thermal relaxation problem by using a much larger number of particles. They used 4096 particles per galaxy, which is sufficiently large to keep the duration of the whole simulation to be a small fraction of the relaxation time of the individual galaxies.

The galaxies gain the thermal energy through encounters. This gain in energy has two outcomes: first, the galaxy as a whole expands, resulting in a *reduction* of the velocity dispersion; secondly, some of the stars which acquire sufficient energy escape from the galaxies. Funato and Makino measured these changes for a wide range of the collision parameters (impact parameter and relative velocity). Their main findings can be summarized as follows. First, the

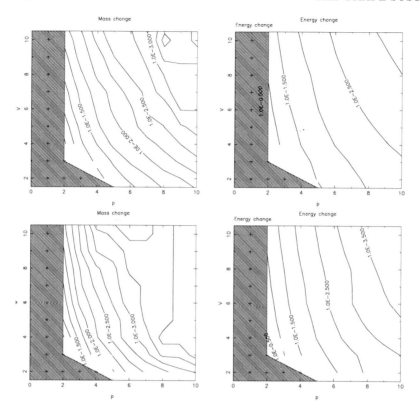

Figure 94 The change of mass and binding energy of galaxies due to an
encounter.

dependence of the change of the mass on the initial structure of the galaxy
is very strong. They used the Plummer model ($\rho \propto r^{-5}$ for large r) and
Hernquist model ($\rho \propto r^{-4}$ for large r). The change in the velocity dispersion
was not much different.

However, the change in the mass was quite different. Figure 94 shows the
contour. We can see that the mass loss of the Hernquist model is much larger
than that of the Plummer model. This result is not really new. Aguilar and
White [AW85] found a similar result, by comparing what they called the
Binney isotropic model and the King model.

Funato and Makino [FM97b] gave a qualitative explanation based on impulse approximation. Unlike the case of merging, here impulse approximation
is alright since the majority of encounters have a very high relative velocity.

They found that the mass loss comes mainly from the stars around the trajectory of the center of mass of the other galaxy. These stars have the largest energy change, and therefore have the largest chance to escape. The energy change, however, is distributed over the whole cluster (except for the very central region, where the impulse approximation breaks down). Thus, the mass loss depends upon the structure of the galaxy more strongly than the energy change. If there is a greater mass at a relatively large distance from the center, the mass loss is larger.

They then integrated the mass loss and energy loss over all encounters, assuming the Maxwellian velocity distribution. Their result indicates that in the case of the Hernquist model, galaxies evolve on the line which is consistent with the Faber–Jackson relation. Note that the Hernquist-like profile itself is likely to develop through interactions [Jaf83], [MAS90].

If the initial distribution of elliptical galaxies is concentrated in ae relatively small area in the mass-velocity dispersion plane, the evolution through encounters establishes the Faber-Jackson relation. In other words, the Faber-Jackson relation might be formed through the evolution of galaxies within the cluster, and therefore is not necessarily universal. The coefficient of proportionality might be different for different clusters.

It might not be too clear for readers whether this result demonstrates anything about the age of the universe. The main conclusion is that Faber-Jackson relation itself tells us rather little about the formation history of the elliptical galaxies in clusters, since it is likely to have developed after the cluster was formed.

6.4.2   Violent relaxation

Funato et al. [FME92b], [FME92a] studied how so-called "violent relaxation" [Lyn67] works in the case of merging and non-merging encounters of two spherical galaxies. Lynden-Bell introduced the concept of the violent relaxation to explain the similarity among elliptical galaxies. At that time, it was well known that the luminosity profiles of most ellipticals were very well expressed by de Vaucouleurs' law (see, for instance, Binney and Tremaine [BT87]), which seemed to suggest that some relaxation mechanism existed.

When a stellar system is in dynamical equilibrium, it can relax only through two-body relaxation. However, during the period of formation through collapse or merging, the system is not in dynamical equilibrium, and particles can change their energies through interaction with the time-varying potential field. Lynden-Bell assumed that this interaction is efficient enough to let the system achieve some kind of statistical equilibrium, and obtained the distribution function of this equilibrium.

A number of researchers have studied the violent relaxation process in one-dimensional mass-sheet model (see, for example, Tanekusa [Tan87] and the references therein) and in three-dimensions (see, for example, van Albada [Van82] and the references therein). Their main conclusion is essentially that the Lynden-Bell distribution is not generally realized.

Funato *et al.* [FME92b] analyzed the way in which the energies of particles are changed by interaction with the time-varying potential field. They considered encounters of galaxies as examples. Their main conclusions are summarized as follows:

(a) In both the merging and non-merging encounters, particles in the central region experience quite a small change in the energy. On the other hand, particles around the half-mass radius experience the largest change in energy.

(b) The change in the energy stops rather quickly, and the average net change in the binding energy of particles in the system is a small fraction of the total potential depth.

Point (a) directly explains why the Lynden-Bell distribution cannot be realized. If some part of the system is essentially untouched, there is no way in which to achieve any statistical equilibrium. The system is far from ergodic. Point (b) explains why the final state depends strongly upon the initial condition.

Heggie [Heg89] reported that the radius of Lagrangian mass shell exhibits persistent oscillatory behavior in the timescale of the local crossing time. Similar oscillatory behavior was also reported in the simulation of cold collapse using direct integration of one-dimensional collisionless Boltzmann equation (for modern, accurate calculation, see, for example, Hozumi and Hernquist [HH95]). Such oscillation, in principle, can affect the energy of particles in the system, and let the system relax.

Funato *et al.* [FME92a] investigated the nature of this oscillatory behavior, using both self-consistent N-body simulation and test particles under fixed potential. With the help of large-N simulation made possible by GRAPE-3, they demonstrated that the persistent oscillatory behavior is explained simply as a statistical fluctuation, and there is nothing mysterious about that.

In their simulation, the amplitude of the oscillation first decays as $1/t$, where t is the time. However, for very large t the amplitude does not vanish. To make matters more confusing, the amplitude itself shows complex oscillatory behavior. This was a rather intriguing result, which seemingly suggested that there might be some mechanism which maintains the oscillation.

They then performed essentially the same calculations but without self gravity, and also found the same complex behavior. Thus, it was clear that there cannot be any physical mechanism that maintains the oscillation, since the particles do not interact with each other in this calculation.

They showed that both the amplitude of the oscillation itself and the time variation of the amplitude can be explained as a statistical fluctuation due to a finite number of particles. Apparent oscillation takes place, since all particles have a similar orbital timescale. Complex changes in amplitude also occur because the timescales of particles are similar but not identical.

In self-consistent simulations, the energy change of particles agrees perfectly with the prediction from two-body relaxation. Sweatman [Swe93] and Kandrup et al. [KMS93] reached essentially the same conclusion as obtained by the two papers by Funato et al. [FME92b], [FME92a].

6.4.3 Groups and clusters

Athanassoula et al. [APB97a] performed a large number of simulations of a compact group of galaxies to investigate the dependence of the lifetime of the group on various parameters, such as the distribution of galaxies and dark matter, the size of galaxies etc. Compact groups are quite strange objects, since it looks unlikely that so many of them exist and currently observable. Hickson [Hic82] catalogued 100 very compact groups of galaxies. Since then, a number of theories have been proposed concerning whether such compact groups are a physical reality or simply the random fluctuations of a projected distribution of galaxies (see, for instance, Hernquist et al. [HKW95] and the references therein). Though a fraction of the compact groups are most likely projection effects, a significant fraction seem to be real, gravitationally bound systems.

The problem with these physically bound groups is that they are so compact that their lifetime must be quite short, and therefore it is unlikely that such systems can be observed. Athanassoula et al. followed the evolution of compact groups from various initial conditions, and found that if the compact group has a massive common dark halo, the lifetime of the group can be significantly longer than that of groups without the common halo.

The velocity dispersion of galaxies in the group is relatively high. This high velocity dispersion implies that most of the mass of the group is in dark matter. In the case of galaxies in the field, the dark matter supposedly forms halos around individual galaxies, as suggested by the dependence of the rotation velocity to the distance from the center of a galaxy. In many spiral galaxies, the rotation velocity of stars and gas in the disk is almost constant for a very wide range in radius, far outside the edge of the stellar disk. If galaxies in compact groups are accompanied by such massive and large halos, they are practically overlapping with each other, and should merge very quickly.

On the other hand, it is also possible that in compact groups, the dark matter is not associated with individual galaxies, but distributed throughout the group. In this case, the lifetime of the group should be longer, since merging cross-section of galaxies is much smaller than the case of individual halos.

The net effect of the presence of a halo, either common (distributed) or individual, is quite complex, since several competing physical effects are working simultaneously. In the case of the individual halo, a larger halo naturally increases the physical collision cross-section. However, a larger mass increases the relative velocity, which decreases the normalized cross-section (see the previous section). In the case of the common halo, a massive halo increases the relative velocities of galaxies and thus decreases the merging cross-section. On the other hand, dynamical friction from the dark halo would cause individual galaxies to sink towards the bottom of the potential well, thereby increasing the concentration of the galaxies.

To obtain a quantitative estimate, therefore, we need to perform extensive N-body simulations. Athanassoula et $al.$ performed a systematic survey of the evolution of compact groups with a wide variety of initial structures. Their main conclusion is that the presence of the common dark halo can actually extend the lifetime of the group quite significantly, in particular if the halo is not very centrally concentrated.

Funato et $al.$ [FME93] performed simulations of clusters of galaxies to study the evolution of galaxies within clusters. As the initial model, they consider virialized clusters composed of identical galaxies or two types of galaxies with different mass and velocity dispersion. The total number of particles was 32 768 or 65 536, and calculation was performed on GRAPE-3. We believe this is the first paper which deals with the evolution of a cluster of galaxies by fully-consistent N-body simulation, in which each galaxy is represented by an N-body system. Previous studies on the evolution of the clusters are based on either Monte-Carlo simulation or the Fokker–Planck approach. The effect of inelastic collision of two galaxies was either approximated by tidal impulse approximation, or the result of N-body simulation was used. As described in Section 6.4.1, the result of N-body simulation was seriously flawed due to the limitation in the number of particles employed, and tidal impulse approximation cannot be applied to low-velocity collisions which cause a merging of two galaxies.

The main result of the N-body simulation is that the galaxies change their structure in such a way that their distribution satisfies the Faber-Jackson relation, if we assume that M/L of elliptical galaxies in clusters is independent of the mass of galaxies in the cluster. On the other hand, the merging and formation of the massive galaxy is quite rare event. The Faber-Jackson relation develops for all simulations, regardless of the initial conditions, such as the initial distribution of galaxies in the cluster or the size distribution of galaxies. On the other hand, the number of merging events depends somewhat upon the initial conditions.

These findings led us to a more detailed investigation of the encounter of two galaxies, discussed in Section 6.4.1.

6.4.4 Algorithm and implementation

Interactions between galaxies and the evolution of a cluster of galaxies are the easiest to simulate, as far as the complexity of the algorithm is concerned. The standard Barnes–Hut treecode with a shared timestep works fine.

In these studies, the preparation of the initial models and analysis of the result require much more programming effort than the time integration itself, though most of the computer time goes on the time integration.

We have been extensively using the NEMO software package, originally developed by Josh Barnes and Peter Teuben (then both were at the Institute for Advanced Study), and now maintained by Peter Teuben [Teu95]. It is designed around the standard data format for N-body (and other) data, which made it possible to develop and maintain a number of tools that generate/modify/analyze N-body data.

6.5   Galaxy formation

How are galaxies and other stellar systems formed? As we discussed in some detail in Section 6.4.1, statistical properties of galaxies give clues to the origin of the universe. However, to use the galaxies as clue, we need to know the formation and evolution history of the galaxies. We already gave an example in Section 6.4.1, but here it would be instructive to give another.

If we can assume that the spatial distribution of galaxies in the universe is uniform and the luminosity of galaxies does not change in time, we can determine the mass (average density) of the universe from the observed luminosity distribution of galaxies. Depending on the mass density, there are three possibilities for the past and future of the universe. If the density is below the critical density, the universe continues to expand indefinitely. If the density is higher than the critical density, the whole universe will reach maximum expansion at a certain time, and will then start to collapse. If the density is just the critical density, the universe would expand indefinitely, but the speed of expansion approaches zero. Note that the density changes in time, and therefore the critical density also changes. In other words, the critical density is a function of the Hubble constant, and we usually measure the density in terms of this critical density. This non-dimensional quantity is usually called Ω.

In terms of Newtonian mechanics, these behaviours correspond to the dynamics of the sphere of uniform density. If the kinetic energy is larger than the gravitational energy, the sphere would expand indefinitely, and if the kinetic energy is smaller, the sphere would eventually recollapse. In this case, the time evolution of the distance between two points can be expressed as the solution of the one-dimensional Kepler problem.

If we know the value of Ω, we can determine the time evolution of the density and Hubble constant, and through this information, we can construct the luminosity distribution of galaxies. Thus, we can determine the value of Ω from the observed luminosity distribution of galaxies by solving the inverse problem. This exercise, which is usually referred to as the "galaxy number count", is an integral part of observational cosmology.

In practice, we have to apply a number of corrections to this idea of galaxy number count in order to obtain a reliable estimate for Ω. First, there is no guarantee that the luminosities of galaxies do not change over time. Second, there is no guarantee that the number of galaxies does not change over time. Thus, we need to evaluate these effects. The formation and evolution of stars in individual galaxies changes the luminosity of galaxies. The merging of small galaxies has changed the number of galaxies in the past. The mergings also change the distribution of luminosities.

The simulation of the formation of galaxies can, in theory, provide the necessary information concerning the evolution of the luminosity and number of galaxies. To model the luminosity evolution, however, we have to simulate the formation and evolution of stars in the galaxy (chemo-dynamics of the galaxy), which will not be covered in this book. Here, we describe some work done using GRAPE hardware.

6.5.1   Purely gravitational simulations

According to present standard cosmology, most of the mass of the universe is in "dark matter". Thus, if our main interest is in the number and mass of the galaxies, we can obtain most of the information from N-body simulations of dark matter. A number of such works have been carried out in the last two decades.

In present standard cosmology, structures such as the galaxies, clusters of galaxies and superclusters are all formed through gravitational instability. When the universe was dense and therefore hot, it was close to thermal equilibrium (isothermal and homogeneous). As the universe expands, it cools down, and fluctuations start to grow.

While fluctuations are small, we can apply the linearized equation and obtain an analytic solution. For the later stage, however, we have to solve the N-body problem.

Fukushige and Makino [FM97a] performed an N-body simulation similar to that of Dubinski and Carlberg [DC91], but with a much larger number of particles and smaller softening. To use the small softening (140pc compared to 1.8kpc of Dubinski and Carlberg [DC91]), they used the individual timestep algorithm with direct summation, which limited the number of available particles to be less than 10^6. Even so, the calculations by Fukushige and Makino [FM97a] are by far the largest for this type of simulation. Most of the runs by Dubinski and Carlberg [DC91] used only 32 768 particles. Navarro *et al.*

[NFW96] performed a large number of simulations, but using less than 32 768 particles.

The result of large-N simulation by Fukushige and Makino [FM97a] was quite different from the results of earlier simulations with a smaller number of particles. The main difference lies in the structure of the central region. Previous works have "shown" that the structure of the dark matter (or "dark halo", as it usually called) is expressed quite well by the Hernquist model, or a modified version of it:

$$\rho = \frac{\rho_0}{r(r + r_0)^\gamma}, \tag{6.35}$$

where $\gamma = 2 \sim 3$. Navarro *et al.* [NFW96] argued that this profile is "universal", in the sense that it did not depend upon the power spectrum of the initial density perturbation or the size of the cluster. If true, this result is quite interesting, and could serve as a practical alternative to "violent relaxation", which was introduced to explain the similarity between elliptical galaxies ([Lyn67]; see Section 6.4.1).

Fukushige and Makino found that the inner region was better expressed by a much steeper cusp ($\rho \sim r^{-1.5}$) for their 768k-particle simulation. To investigate the cause of the difference, they also performed calculations with larger softening and a smaller number of particles. For these calculations, they found a result similar to that of Navarro *et al.* [NFW96]. The difference in the power index at the central region is mainly due to the difference in the number of particles.

The central cusp with a density profile of $\rho \sim r^{-1.5}$ has the velocity dispersion *decreasing* inward. Thus, in the thermal timescale, the thermal energy flows inwards, just as in the case of the gravothermal expansion phase of the gravothermal oscillation, and the central density decreases. This gravothermal expansion is the main reason why simulations with a small number of particles resulted in the cusp being much shallower than that obtained by simulations with a large number of particles. Whether or not the result obtained by Fukushige and Makino is "universal" in the sense Navarro *et al.* claimed remains to be seen. At present, the calculation performed by Fukushige and Makino is quite expensive, requiring weeks of CPU time on a fully configured GRAPE-4. Thus, it is still not feasible to run many models.

6.5.2 The effects of hydrodynamics and other physical processes

In the formation of galaxies, the role of the gas dynamics and star formation process is certainly important. If there are no gases in the system, the stellar disks will never be formed. However, our group has not done any studies on a galaxy in which gas dynamics and the stellar formation process are included.

On the other hand, almost all researchers who are using GRAPE hardware outside our group are doing gas dynamics simulation, in relation to the formation of galaxies and other stellar systems. In this section, we briefly overview this work. Thus, this section is not really limited to the formation of galaxies, but includes the formation and evolution of a wider range of astronomical objects.

Umemura et al. [UFM+93] were the first to implement a gas dynamics simulation using SPH on GRAPE hardware (GRAPE-1A). This code was used to study the possibility of the direct formation of massive black holes from the gravitational collapse of primordial density perturbation [UFM+93]. They concluded that massive black holes could be formed directly from the collapse of primordial density fluctuation, if the Compton drag from the cosmic background radiation is efficient enough.

Steinmetz [Ste96] described in detail his implementation of SPH on a GRAPE-3A system. Steinmetz and Müller [SM94], [SM95] performed the simulation of galaxy formation using this code. In their simulation, the initial condition is a sphere with uniform density which rotates rigidly. Besides gravitational interaction, the following physical effects and processes are included:

- gas dynamics
- radiative cooling of gas
- star formation from "converging" region
- stellar evolution and supernova explosion.

In more recent papers, Steinmetz and his collaborators proceeded further to include the effect of the UV background radiation. Here, we briefly describe the simulation performed by Navarro and Steinmetz [NS97]. In this particular study, they did not include the star formation.

For the initial condition, they used the hierarchical technique as used in Navarro and White [NW93]. The basic idea of this technique is to perform the simulation of a large scale in low resolution, and then identify the regions of interest (where galaxies are likely to be formed), and perform the simulation of these regions with a higher resolution. The initial simulation was done using 64^3 particles in a cubic cell of 30Mpc per side. The high-resolution simulation uses 40^3 dark matter particles and the same number of gas particles. The fraction of the gas mass is 5% of the total mass, thus the mass of a gas particle is $1/19$ of that of a dark matter particle. They have clearly demonstrated that the UV background radiation prevents the formation of a compact gas clump.

Mori et al. [MYTN97] performed a simulation of the formation of dwarf galaxies, using a method similar to that described by Steinmetz and Muller [SM95]. They found that the heating by supernova explosions has rather a complex effect on the dynamics of the system. Initially, burst star formation occurs at the central region. This starburst then leads almost simultaneously to supernovae. As a result, a supersonic outflow of gas is formed. This supersonic flow collides with the still-infalling low-temperature gas and compresses it through shock. The high-density region formed by this shock compression then starts to form stars. They argued that this dynamical process is responsible for the formation of an exponential brightness profile.

6.5.3 Algorithm and implementation

Implementing SPH efficiently on GRAPE hardware requires considerations that are quite different from those for implementing the gravity-only calculation codes. The reason is that the calculation of hydrodynamic interactions contain a non-negligible fraction of the calculation cost. The calculation of gravity is greatly accelerated by GRAPE, but that of the hydrodynamic interaction is not. Thus, the calculation of the hydrodynamic interaction must be carefully optimized to achieve a high overall speed.

Basic SPH formalism

The basic idea of Smoothed Particle Hydrodynamics (SPH) is to model the gas as a collection of "particles" or Lagrangian fluid elements, which feels the pressure force as well as the gravitational force. In a typical SPH formulation, a physical quantity A is approximated by a smoothed estimate

$$< A(r) >= \int d^3 A(\mathbf{r}')W(|\mathbf{r} - \mathbf{r}'|, h). \tag{6.36}$$

The above formula is simply the convolution of A and the smoothing kernel function W. We usually choose W with a strong peak at $\mathbf{r} - \mathbf{r}' = 0$, so that it does not oversmooth the original function A. In practice, we know the physical quantities only at the position of particles. Thus, the integration in Equation (6.36) is replaced by a summation

$$< A(r) >= \sum_j m \frac{A(\mathbf{r}_j)W(|\mathbf{r} - \mathbf{r}_j|, h)}{\rho(\mathbf{r}_j)}. \tag{6.37}$$

The most widely used form of the kernel function W is the so-called "spline kernel"

$$W(r, h) = \frac{8}{\pi h^3} \begin{cases} 1 - 6u^2 + 6u^3, & 0 \le u \le 1/2 \\ 2(1 - u)^3, & 1/2 \le u \le 1 \\ 0, & 1 \le u \end{cases} \tag{6.38}$$

where $u = r/h$. This kernel has many convenient characteristics. First, it has a compact support. Thus, the contributions from particles outside this compact support are exactly zero and need not be calculated at all. Second, the polynomial form is inexpensive to evaluate.

To implement an efficient and accurate SPH algorithm, we have to consider a number of details, such as how to vary the smoothing length h depending on the local density, how to implement artificial viscosity, how to implement an individual timestep and so on. Here, we will not go into these technical details. Instead, we briefly discuss the practical considerations of using GRAPE hardware with SPH.

So far, GRAPE systems have been used for two different purposes in SPH simulations. The first is to calculate the gravitational interaction, and the second is to construct the list of neighbor particles. In the simplest case, all particles share the same timestep, and GRAPE is used to calculate the gravitational force and the neighbor list at each timestep. In this case, GRAPE can accelerate the overall calculation speed quite significantly.

For the collapse simulation with 10k dark matter particles and 10k gas particles, Umemura *et al.* [UFM+93] reported a speedup of about a factor of 100, for the combination of a Sun SPARCstation 1 and GRAPE-1A over the calculation on a SPARCstation 1 only. This very impressive result, however, is somewhat unfair, since the algorithm used for the calculation on the general-purpose computer was suboptimal. In particular, the simple direct summation was used to calculate the gravitational force. For the total of 20k particles, the Barnes–Hut treecode would be 20–30 times faster than the direct summation (see, for example, Hernquist [Her87]). The calculation cost of constructing the neighbor list is comparable to that of the force calculation itself, in the case of the treecode [HK89]. On the other hand, with direct summation, the neighbor list is constructed for free (since the force calculation is so much more expensive). Thus, the actual gain by treecode is more like a factor of 10. Even so, compared to the treecode, the relative performance gain of the SPH calculation using GRAPE is roughly a factor of 10.

The reason why the speedup achieved using GRAPE is limited for SPH is quite simple. If one use the treecode, the number of particles and nodes one particle interacts with is around 300 or so. This number depends both upon the total number of particles and the required accuracy, but 300 is a typical number.

On the other hand, the number of particles in the SPH kernel is 30–60, depending on the implementations. The number of floating-point operations per pairwise interaction is comparable to a gravity calculation in treecode and the calculation of a SPH interaction, though SPH interaction is somewhat more costly. Thus, calculation of SPH interaction occupies around 10% of the total calculation cost. Amdahl's law tells us that we cannot get more than 10 times speedup, even if we have infinitely fast GRAPE hardware.

Individual timestep

A similar result has been obtained by Steinmetz, who implemented SPH simulation code on a GRAPE-3A system. He called his code just "GRAPESPH". It calculates the gravity directly. The main difference from Umemura's code is that it implements the individual timestep in a way similar to that described in Hernquist and Katz [HK89]. In Steinmetz's implementation, however, both the calculations of SPH interaction and that of gravitational interaction are done utilizing the individual timestep, while in Hernquist and Katz [HK89] only the gravitational force calculation was done using the individual timestep. SPH interaction is always calculated for all particles at each smallest timestep in Hernquist and Katz [HK89]. This made sense, since gravity is more than a factor of 10 expensive on a conventional computer. However, with GRAPE, SPH force must be done individually, since it consumes a comparable or more CPU time than gravity.

In practice, with the individual timestep, the gain by using GRAPE for gravity is larger than that for a shared-timestep implementation. This is simply because the present individual-timestep SPH codes typically have a rather large overhead when the number of particles to be integrated is a small fraction of the total number of particles.

The individual timestep algorithm used here is essentially the same as the hierarchical timestep algorithm discussed in Section 5.1.3. At each timestep, we need to predict the position of all particles, or at least that of all nodes and particles that exert forces on particles to be integrated. In TREESPH of Hernquist and Katz [HK89], the tree is constructed from scratch at each timestep. This implies that even if only a few particles are to be integrated, the calculation cost is $O(N \log N)$.

In GRAPESPH of Steinmetz [Ste96], the positions of all particles at the current time are predicted on the host computer and sent to GRAPE at each timestep. Thus, even if only a few particles are to be integrated, the cost of communication and prediction on the host computer is still $O(N)$.

At present, this overhead does not cause a serious performance problem, simply because the range of the timestep is still fairly narrow (typically less than 1000). However, as we increase the number of particles further, the overhead will become more significant.

The main objective for increasing the number of particles is to achieve better resolution, in other words, to resolve the central, high-density region. We cannot simulate a structure smaller than the smoothing length. Even if we use adaptive and variable smoothing, we cannot study the structure of the region whose mass is not much larger than the mass of a single particle.

A larger number of particles thus allows us to study the central region more accurately. However, it implies that the minimum timestep would go further down. Theoretically, if we have a singular isothermal cusp, the ratio between the average timestep and minimum timestep is $O(N^{2/3})$ [MH88].

Note that this is the limitation of current implementation of the time integration on a particular GRAPE hardware, and not a fundamental limitation. In the case of GRAPE hardware, this overhead can be eliminated by performing the prediction on GRAPE, as in the case of the GRAPE-4 system. In the case of the pure TREESPH, we could in principle implement the local tree reconstruction in the way discussed in McMillan and Aarseth [MA93a].

Periodic boundary

To perform a realistic simulation of galaxy formation in a cosmological context, it is desirable to implement a periodic boundary condition, or at least to include the tidal effect from the outside of the system in some way.

Brieu et al. [BSO95] implemented the P^3M algorithm on GRAPE-3A hardware. The P^3M algorithm works in roughly the same way as the Ewald method. The only difference is that the Ewald method uses the Discrete Fourier Transform (DFT) to obtain the contribution of long wavelength components, while P^3M relies on FFT. Thus, the calculation cost of the Ewald method is proportional to NM, where M is the number of Fourier components, while that of P^3M is proportional to $N + M \log M$. On the other hand, the error in the force is much larger for P^3M if the same M is used, due to the gridding error.

GRAPE-3A can calculate the force with pure $1/r$ potential only, while P^3M requires the force in the real space to be truncated. Brieu, et al. [BSO95] implemented this truncation by a linear combination of several force terms calculated with different softening lengths.

Klessen [Kle97] proposed another way to implement the periodic boundary, which is based on the idea of "Ewald correction".

Hardware for SPH interaction

As described above, the calculation cost of SPH interaction is typically around 10% of the total calculation cost, if treecode is used to calculate gravity. Thus, even if GRAPE is infinitely fast, the speedup factor cannot exceed 10.

In many situations, this factor of 10 is sufficient to justify the use of GRAPE hardware. As long as the cost of GRAPE hardware is less than 10 times the cost of the host computer, the overall cost performance is better for the GRAPE system. Moreover, the improvement in speed of the host computer directly improves the speed of the overall calculation, since the calculation speed is limited by the speed of the host computer. On the other hand, to achieve the same speed as is available on GRAPE, we need at least a 10-way parallel computer, and we have to develop a parallelizable program for that machine.

If we were to further improve the speed of SPH calculation, we could develop a GRAPE-like hardware for SPH interaction. The calculation of SPH interaction is similar to that of gravity. We calculate the distance between two particles and evaluate some function of this distance and other physical quantities, and then accumulate the contributions from many particles. Thus, it is not unthinkable to use GRAPE-like hardware to evaluate the SPH interaction. Very roughly speaking, we can get another factor of 10 speedup, since the evaluation of SPH interaction occupies around 90% of the calculation cost, if we put aside the calculation of the gravity and construction of the neighbor list.

A group at the National Astronomical Observatory of Japan has been working to develop GRAPE-like hardware for SPH calculation [YOT+96].

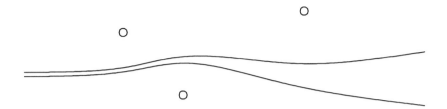

Figure 95   Exponential divergence of two nearby orbits.

6.6   Chaos in N-body system and multiple gravitational lensing

6.6.1   Chaos in N-body system

Before discussing the application of GRAPE hardware to the study of the effect of gravitational lensing, we first describe the theoretical background. The basic concept is very old, but it was not fully understood until very recently.

It is well known that a self-gravitating N-body system is chaotic, in the sense that a slight difference in the initial condition grows exponentially in time. This phenomena was first observed by Miller [Mil64]. For N in the range of 4 to 32, he found that the e-folding time is smaller for a larger number of particles. This fact remained a mystery until the more accurate measurement for runs with a larger number of particles and the theoretical understanding were supplied by Goodman *et al.* [GHH93].

There were a number of studies after the pioneering work by Miller [Mil64]. In particular, Gurzadian and Savvidy [GS86] argued that this process was directly related to a new form of relaxation process, which had a timescale shorter than the usual two-body relaxation. Kandrup [Kan90] also reached a similar conclusion by measuring the curvature of the phase $6N$-dimensional phase space. In several papers which followed (Kandrup *et al.* [KMS94] is the final one), Kandrup and his collaborators performed a number of numerical experiments, and in early works [Kan90] claimed that this process caused "mixing". However, in their last paper [KMS94], they have shown that energy and angular momentum change much more slowly than expected from the timescale of the orbital divergence.

Goodman *et al.* [GHH93] were the first to give a correct theoretical understanding of the problem. Their numerical experiment with up to 256 particles established that the timescale converges to the crossing timescale (around $0.1t_{cr}$) for large N.

The chaos means that the distance between the systems with very close initial conditions increases exponentially in time. As the simplest example, let us consider the orbits of two test particles which move in the potential field of N other point-mass stars placed in fixed locations (see Figure 95). Each time these two particles pass by a star, their orbits are deflected by the gravitational force from the star, and the deflection angles are slightly different due to the difference in the impact parameter. This difference in the deflection angle is proportional to the distance between two orbits, because the difference in the force, which is expressed as the tidal force, is proportional to the distance between two test particles. Therefore, the relative angle between two orbits, as well as the distance between the orbits, grows exponentially. This exponential growth continues as long as the relative distance of two orbits is smaller than the typical impact parameter for encounters which are most effective in increasing the distance. If the relative distance becomes larger than this typical impact parameter, the growth timescale becomes longer. Both the theoretical estimate and numerical experiments have shown that this distance is expressed as

$$b_{crit} \sim RN^{-1/2}, \qquad (6.39)$$

where R is the size of the system and N is the number of particles in the system, if the system is in virial equilibrium.

All of the above studies dealt with the divergence of the trajectory of two neighboring N-body systems. They discussed how two N-body systems of similar initial conditions divert from each other. From the theoretical point of view, this is a very important problem, since it is directly related to the question of the meaning of the N-body simulations.

If the timescale of the exponential divergence is much shorter than the timescale covered by the simulation, it seems to imply that the trajectory of the N-body system is chaotic, and that the states of two N-body systems initially close to each other would very soon become totally different from each other.

This naïve interpretation of the exponential divergence, however, seriously contradicts the traditional picture of relaxation in the N-body system, which we described in detail in Section 6.2. In the standard theory of two-body relaxation, the distribution function would change only in the relaxation timescale, which is longer for larger N. On the other hand, the timescale of exponential divergence is a constant fraction of the crossing time, independent of the total number of particles N. Thus, at first sight, it seems that the theory of two-body relaxation tells us that the system relaxes in the thermal timescale, while the theory of exponential divergence tells us that the system should forget its initial condition in a few crossing times. Since the relaxation implies that each particle "forgets" its initial condition, these two theories seem to contradict.

Goodman *et al.* did solve this problem. For the orbits of a pair of particles, the exponential growth slows down when the distance between the two orbits reaches b_{crit}. While the separation is smaller than b_{crit}, close encounters with other particles take place coherently for the two orbits, and this coherence is the condition of the exponential growth. Once the distance has reached b_{crit}, two particles experience independent close encounters and the distance increases through independent random walks. In other words, the distance grows as $t^{1/2}$. Their argument of this slowing down is based on purely analytical treatment.

Fukushige *et al.* [FMNE95] performed a simulation of the exponential divergence of two orbits in a fixed distribution of particles, using up to 8192 particles. They clearly demonstrated that the slowing down of the growth occurred at the distance b_{crit}. The large number of particles allowed them to measure the point of saturation and its dependence on N much more accurately than was done in previous studies.

From the theoretical point of view, this result is quite important. It clearly demonstrated that the exponential growth *does not* lead to relaxation, because it saturates at a distance smaller than the system size by a factor of $1/\sqrt{N}$. After the distance of the two orbits reaches this critical separation, the distance grows following a random walk, in other words, through the usual two-body relaxation. In the limit of $N \to \infty$, the length scale for which the exponential divergence takes place becomes zero.

Even though the exponential divergence does not lead to relaxation, it still means that the positions of particles after a few crossing times is not informative. Thus, it would be useful if there were some way to extend the timescale of the exponential divergence.

Goodman *et al.* suggested that the timescale should be proportional to $\epsilon^{2/3}$, if $\epsilon > b_{crit}$, using rather complex theory. The result of their numerical experiment was, at best, not inconsistent with the theory. Huang *et al.* [HDC93] performed a similar study, and found that the power index is noticeably lower than 2/3 (around 0.45). Their numerical result showed a clear indication that dependence is weaker for larger values of softening.

Fukushige *et al.* [FMNE95] measured the dependence of the timescale on the size and shape of the stars. They obtained a power index noticeably larger than 2/3, but this is partly because of the way in which they calculated the timescale, which systematically overestimated the timescale when it is long. Fukushige [Fuk96] performed a more accurate measurement with an improved calculation method, and obtained a result which is in good agreement with the theoretical estimate of Goodman *et al.*.

Thus, we can extend the timescale of the exponential divergence quite significantly by setting the softening to be larger than b_{crit}. If we choose softening to be comparable to the average interparticle distance, the timescale is proportional to $N^{1/9}$. If we choose the "optimal" softening of $N^{-1/5}$[Mer96], the timescale is $N^{1/5}$.

This, of course, does not mean that the two-body relaxation effect is suppressed, since the two-body relaxation effect is reduced only through the change in the lower limit for the Coulomb logarithm. In particular, if the softening is chosen to be b_{crit}, the relaxation timescale is roughly a factor of two longer than that for point-mass particles.

6.6.2   Multiple gravitational lensing

The exponential growth does not change the relaxation timescale of the collisional system and has no effect on the evolution of the collisionless system. So does this effect have any astrophysical significance other than as the theoretical basis for the validity of N-body simulation? The answer seems to be "yes", but not for the evolution of stellar systems.

Recently, gravitational lensing has become to be used as the probe to the mass distribution in various scales. In the smallest scale, the microlensing events of the stars in Magellanic clouds are used to estimate the mass distribution in the halo of our galaxy, and in the largest scale, the lensing of QSOs are used as the test for cosmological models. There are numerous works which used the lensed images of background galaxies as the probe to the mass distribution within clusters of galaxies. Along with the direct measurement of the luminosity and power spectrum of X-ray emission, gravitational lensing is now widely used to measure the mass of a cluster of galaxies. Since there are so many background galaxies, in many cases we can obtain very good statistics now that very high-resolution imaging has become possible with HST.

In all these studies, however, the lensing is either modeled as a single event or as a linear combination of independent deflections by single lensing objects. In other words, the effect of exponential divergence discussed in the previous section is neglected.

This treatment is perfectly valid for applications such as the microlensing due to halo stars, because the optical depth of the halo is so small (order of 10^{-6}). However, in the cosmological scale, the effect can be very large. If the universe is flat, the timescale of the exponential divergence is a small fraction of the Hubble time, since the Hubble time is equal to the dynamical timescale of the whole universe. Thus, in theory, the trajectory of the photons from cosmological distance could be affected by this exponential divergence.

In particular, the angular correlation of the Cosmic Microwave Background Radiation (CMBR) could be strongly affected by the lensing in a wide range of angular scales, depending on the cosmology. There were many theoretical estimates of the effect of the gravitational lensing on the angular correlation of CMBR, and almost all of them concluded that the effect is fairly small (see, for example, Seljak [Sel96] for a recent estimate). However, these estimate were based on the assumption that the effect of multiple lenses can be linearly added. In other words, the effect of the exponential growth is ignored.

Fukushige et al. ([FMNE95]) performed extensive simulations to evaluate the effect of the exponential growth, and found that the effect is much larger than that of the previous estimate, at least in small scales (around arcminutes), where the inhomogeneity is highly developed. Whether or not it has any significant effect on larger scales, such as the fluctuation in 10-degree scale measured by COBE (see, for example, Banday *et al.* [BGT+94]), depends very strongly on cosmological parameters such as Ω and the actual degree of the inhomogeneity. It should be noted that the estimate by Fukushige *et al.* [FMNE95] was based on the *wrong* estimate of the dependence of the timescale on the softening (the size of the lensing objects), which caused an *underestimate* of the effect of physically extended objects.

The anisotropy of CMBR is not the only possible application of this exponential growth. Funato *et al.* [FME94] performed a simplified calculation to evaluate the effect of the exponential growth on the luminosity of high redshift objects, and found that it can have a noticeable effect on the galaxy number count statistics.

The gravitational lensing does not change the average luminosity, but it amplifies a small fraction of a distant object very strongly. Thus, the luminosities of other objects are, on average, attenuated, to compensate for the strong amplification of a small fraction of the total objects. The median attenuation effect in magnitude is expressed as

$$\Delta m \simeq 3\sqrt{\Omega_l}\log(1+z) \qquad (6.40)$$

in the case of a flat $\Omega = 1$ universe. Here, z is the redshift and Ω_l is the fraction of the total mass of the universe which contributes to the exponential growth due to lensing. It could be anywhere in between 0.01 and 1, but even with a rather modest assumption of $\Omega_l = 0.04$, the median attenuation at $z = 2$ is more than one magnitude.

An analytic estimate of the attenuation effect based on the assumption of linear combination of multiple lensing would give the average attenuation as a linear function of the optical depth, not the exponential function. In other words, the attenuation would be log-log of z, not $\log z$ as given in Equation (6.40).

To summarize this section, we have made very preliminary studies on the effect of the exponential growth caused by the multiple gravitational lensing on CMBR anisotropy and luminosity of a high-z objects. Our result indicates that the effect is much larger than previous estimates, which neglected the exponential divergence. We need to perform more detailed and realistic simulations, taking into account the structure of the lensing objects, the spatial distribution, the the effect of cosmological parameters such as Ω and Λ, to quantitatively discuss the effect.

7

The Future of Special-Purpose Computers

We believe our GRAPE project has so far been quite successful. In this chapter, we try to predict the future of GRAPE and other special-purpose computers. We first discuss earlier special-purpose machines, in particular Ising machines and QCD machines, to see how the evolution of machines was influenced by technological issues. Then we discuss whether a GRAPE-type architecture will be competitive in the future, and finally whether special-purpose computers will be used in other fields of computational science.

7.1 Looking back: Ising machines and QCD machines

The history of other special-purpose computers, in particular the Ising machines and QCD machines, teaches us a lot about how a particular type of special-purpose computer evolves.

Ising machines adopted hardwired pipelines as the basic architecture of the machine, while QCD machines used microprocessors as the basic building blocks. These were quite natural choices. The development of Ising machines started in the early 1970s and that of QCD machines in the early 1980s. Microprocessors with a reasonable arithmetic speed became available only in the early 1980s. On the other hand, small-scale IC chips became available in the late 1960s. In the 1970s, it was not very difficult to design and build a single pipelined processor specialized to a certain application using off-the-shelf ICs. This is especially true for Ising machines, which required a very small amount of logic to implement the hardwired pipeline to flip spins. The reason why Ising machines were more cost-effective than general-purpose computers is essentially that Ising machines required a much smaller amount of hardware than was necessary for general-purpose computers with the same speed.

However, if we follow the history of Ising machines, we will find that this advantage of the Ising model quickly changed to a disadvantage, in that it was impossible to utilize the large number of transistors which became available

Table 9 Technology evolution and special-purpose computers.

	1970s	1980s	1990s
Technology	IC	Microprocessors	VLSI ASICs
Ising machines	good	small gain	—
QCD machines	hard	good	small gain
GRAPE	very hard	hard	good

as the device technology advanced. With an Ising spin calculation, a spin is fetched from memory, applied with a few operations and stored back to the memory. Thus, the required memory bandwidth to support the processor is pretty high. In the 1970s, reasonably high memory bandwidth was not too difficult to achieve, since the memory access time was not much longer than the typical logic delay time. However, it became increasingly more difficult to supply the necessary bandwidth as LSI technology advanced. This is similar to the problem with microprocessors which we discussed in Chapter 2, but much more severe because the number of transistors needed to construct an Ising processor is far smaller than the number of transistors needed to construct a microprocessor with a decent floating-point performance.

The QCD machines based on microprocessors did provide a cost performance that is significantly better than that of vector machines or minicomputers in the 1980s. As we have seen, this performance advantage was realized using many microprocessors in parallel. To put it in another way, the success of QCD machines was because they used CMOS technology which offered a much larger number of transistors than the technology used in vector machines.

This advantage has almost vanished with the advent of commercial MPPs and SMP clusters, which are in many cases direct descendants of QCD machines. It is not impossible to get an advantage in performance of something like a factor of five or ten by optimizing the word size and communication network. However, the relative advantage is clearly smaller than what was possible 10 years ago. Table 9 gives a summary of the above argument. Ising machines are best suited for implementation using SSI, which was the most advanced technology of the early 1970s. Thus, the relative advantage of Ising machines over general-purpose computers was largest at that time. For QCD machines, which were implemented using off-the-shelf microprocessors, the best time was the mid-1980s.

These two examples clearly demonstrate that an architecture which was good at one time ceased to be so in a rather short time. This change took place because of advances in device technology. In a wildly simplified sense, Ising machines were made possible by SSI, which integrated a small number of gates in one chip, and become less competitive as the gate count on one chip continued to increase.

Table 10 The memory bandwidth relative to the computing speed.

Machines	Bytes/flops
GRAPE-1	0.2
GRAPE-2	0.8
GRAPE-3	0.014
GRAPE-4	0.014

The advantage of the QCD machines stemmed from the fact that the designers of QCD machines took advantage of CMOS device technology, or more specifically, one-chip microprocessors, before the designers of general-purpose high-performance computers did. Thus, once general-purpose high-performance computers also switched to CMOS microprocessors, QCD machines lost their architectural advantage.

In both cases, the bottleneck that limits the relative advantage of the special-purpose approach is the chip-to-chip communication bandwidth.

7.2 The advantage of GRAPE systems

The GRAPE architecture has two significant advantages. The first is that the required memory bandwidth is extremely low. The second is that the hybrid architecture of the general-purpose host and the special-purpose GRAPE simplifies the development of both the hardware and software. Let us analyze these two points in more detail.

The first advantage is the low requirement for memory bandwidth. Even in the case of a single pipeline, about 30 operations are performed per three words of input data, and no output is necessary during the force calculation. Moreover, many pipelines can share the same input data. In the first place, this low requirement for memory bandwidth made it possible to pack many arithmetic units onto a single chip without hitting the pin count limit. It also makes it possible to use a number of pipelines in parallel.

Thus, the GRAPE architecture effectively solved the chip-to-chip bandwidth problem, first by constructing a pipeline of many arithmetic units that requires low input bandwidth, and then allowing many such pipelines to share the same input data stream. Table 10 shows the memory bandwidth of GRAPE systems relative to their computing speed. Compared with the numbers in Table 1, these numbers are smaller by several orders of magnitude. Note that in Table 1 we listed the number in word/flops, where one word is typically 8 bytes.

This characteristic of the GRAPE architecture comes from the nature of the application. GRAPE deals with systems of particles with a long-range interaction. As a result, the amount of computation per particle is very large. This is true even if we use sophisticated algorithms like FMM or Barnes–Hut tree. In addition, a high level of parallelism is achieved with low memory bandwidth, since all particles feel the forces from all other particles.

The second advantage comes from the hybrid architecture. As we stated earlier, there is an important difference between GRAPE systems and other systems developed for particle simulations, or other special-purpose computers. GRAPE systems are designed to work as an attached processor which performs only the force calculation part of the simulation. Everything else is done on the front-end computer. This approach has several important advantages over other approaches. The most important one is that it makes the development of hardware much simpler than it would be in the other approaches, simply because the amount of hardware to be designed is smaller. It also reduces the cost of software development, since the software for the front-end computer is already there. The system software, such as the operating system and compilers, is developed and debugged by somebody else. Even the simulation software itself is mostly recycled from what was already developed and used on general-purpose computers.

Another merit of this hybrid approach is that the lifetime of the machine is significantly lengthened by upgrading the host computer. With the GRAPE-type hardware, in many cases the GRAPE processor is initially too fast for a host computer with an affordable price tag. Thus, for a few years we can get significant performance improvement just by upgrading the host computer. This is actually a regular exercise in many institutes which currently own GRAPE hardware.

The advantage of our approach is a consequence of the contemporary environment. For example, the strategy of separating the computational work between a general-purpose front-end and a special-purpose attached processor has become a practical solution only after microprocessor-based systems such as personal computers and workstations have become competitive as a platform to perform number crunching. In the 1970s they did not exist. A hybrid machine was not a viable strategy, since even a minicomputer was still too expensive to be used as a dedicated front-end computer. Therefore, even though almost all machines had a "front-end", the task of the front-end computer was limited to things like the setup of the initial condition and retrieval of the final result.

To summarize, the key reasons why GRAPE systems achieved very high performance are the following:

1. The basic strategy of constructing a hybrid computer with a general-purpose front-end and a special-purpose back-end.
2. Use of a fully pipelined processor specialized and optimized to the calculation to be performed on the special-purpose computer.

3. Use of custom LSI technology to implement the pipeline processor, which make it possible to construct a massively parallel system.
4. Choice of the application and algorithm which makes the above three strategies possible.

As stated earlier, the first point is a very environment-dependent one. If personal computers or workstations were not available, our approach would not be possible. Thus, our approach requires a highly developed device technology which allows the mass production of high-performance microprocessors.

The second and third points also depend strongly upon the device technology available. A fully pipelined processor for a specific application requires a very large number of transistors. To assemble a pipeline such as that used in GRAPE-1 in early 1970 would not be an easy task. In 1980, GRAPE-1 with reduced accuracy might not have been very difficult, but GRAPE-2, which utilizes 32- and 64-bit floating-point numbers, would be very difficult. A machine like GRAPE-2 became possible when floating-point chipsets such as the Weitek 1164/1165 appeared, which was used in many early special-purpose and massively-parallel computers.

The use of a custom LSI to implement the pipelined processor become viable only very recently. The HARP chip used in GRAPE-4 integrates some 20 floating-point units, on a chip as large as 200 mm^2 containing 400K transistors. By today's standard this is not a large chip. One can integrate more than ten million transistors on a chip, and typical microprocessors do have more than five million transistors. However, just 10 years ago, the number of transistors that could be integrated on a chip was more like 50 000, which is not sufficient to integrate a single 64-bit floating-point multiplier. Thus, it would have made more sense to use these floating-point LSI chips as building blocks than to develop a custom LSI, since the number of LSI chips needed to construct the system would not be vastly different, and the development cost of custom LSI chips would be high prohibitingly, since several chips must be developed.

The choice of the algorithm is a critical design issue. We will address this problem later.

Therefore whether an approach used in the GRAPE systems will be effective in future depends very much upon how the computer industry and silicon LSI technology evolve.

At the level of the overall architecture, the GRAPE architecture requires low-priced general-purpose computers with a price-performance comparable to that of high-end computers. The present trend in the high performance computer industry seems to guarantee the existence of low-cost computers for the predictable future.

One obvious problem is that microprocessor-based systems do not look like the most cost-effective way of constructing a high-performance general-purpose computer. However, so far no better alternative has yet been invented.

Microprocessor-based systems are not a really good solution because they use only a very small fraction of the total number of transistors available in state-of-the-art LSI devices.

On the other hand, there are many reasons to believe that the present microprocessor-based systems will continue to thrive for the next 10 or 20 years. The first reason is that it will take time for a widely accepted computer architecture to be replaced by something else. An obvious example is the vector processor. Though the CMOS device technology to build fast floating-point units was already available in 1985, vector machines are still on the market offering price-performance not vastly different from that of MPPs. In fact, for many real applications, traditional vector processors are still faster than MPPs, even though the peak speed of the latter is more than 10 times faster than that of the former.

An interesting prediction about how the computer architecture would evolve in the next 20 years can be found in Sterling *et al.* [SMS95]. They considered three different architectures, which they called categories I, II and III. Categories I and II are extrapolations of present-day vector-parallel machines and MPPs, respectively. Category III is the extrapolation of the approach currently realized in the AMD SHARC processor, which integrates a modestly-complex processor with a modestly-large memory. Integration of the memory and processor onto a single chip has an enormous advantage in that it provides extremely high bandwidth between the memory and processor, as long as communication between processors which reside on physically different chips is not very often (for an introductory review of this idea, see Patterson *et al.* [PAC+97]). This architecture already has several different names: PIM (Processor In Memory), IRAM (Intelligent RAM), PPRAM (Parallel Processing RAM).

The PIM approach will have a range of applications even more limited than what present MPPs have achieved because of the extremely limited communication bandwidth and rather limited amount of memory. For applications like Lattice QCD simulation, which requires a relatively small amount of memory, PIM architecture might prove to be the most efficient, but for many other applications it might not be the best.

Silicon VLSI technology will evolve in much the same way as it has done for the last 25 years. The number of transistors will continue to increase. The clock frequency will also evolve, but not as rapidly as it has in the last 10 years. The I/O bandwidth of a chip will increase, but this increase will be much slower than that in the processing power of the chip (see Table 2).

This trend in silicon VLSI technology is ideal for GRAPE. As we stated earlier, the main advantage of the GRAPE system is its exceptionally low requirement of memory bandwidth. Thus, the relative advantage of the GRAPE system over conventional general-purpose computers will continue to increase as long as advances in silicon VLSI technology follow the present trend.

Of course, the relative advantage of GRAPE could vanish if some revolution in the architecture of the general-purpose computer makes it possible to circumvent the limitation in the communication bandwidth. For example, a PIM-type approach can, in principle, allow very efficient use of the transistors, if the requirement for interprocessor communication bandwidth and latency is small. At present, we cannot tell whether or not PIM-type machines will be competitive with the GRAPE-type approach for N-body simulation.

Not every aspect of the development of silicon VLSI technology has a plausible effect on the price-performance of the GRAPE system. The largest problem is the development cost of the LSI. The development cost of the VLSI chip has been increasing, and will continue to increase. The development cost an ASIC manufacturer quotes for the most advanced VLSI chip is currently well over \$ 100 000 and approaching \$ 1 million (early 1997).

There are several reasons for this increase in development cost. As chips become larger, the design becomes more complex. This complexity in design pushes up the development cost. The increase in the capital investment for the fabrication facilities might be reflected in the design cost as well. In any case, if the initial development cost continues to increase, the relative advantage of GRAPE-type machines will diminish. If the total budget we can spend is not sufficient to develop a custom LSI, there is not much chance for the future of GRAPE.

Whether the development costs will become a serious problem or not remains unclear. In theory, the development cost can be reduced by a large factor if several different designs are made from a single wafer. This approach is reportedly used by MOSIS and also internally in the SONY Corporation.

Yet another way to reduce the development cost is to use a "field-programmable" LSI chip for the pipeline processor. This option will be discussed in more detail later.

In summary, it seems fairly reasonable to expect present trends in the evolution of computer and semiconductor device technologies to continue, at least for the next 10 years. This means that the architectural advantage of the GRAPE system will become more pronounced in the foreseeable future.

7.3 GRAPE-6: A petaflops GRAPE

In this section, we describe the design of GRAPE-6. Here we assume that we will start the design very soon, in the 1997–1998 time frame, and will complete the machine by the year 2000. We have just received grant from the Ministry of Education, which starts in the 1997 fiscal year. The size of the machine to be built is limited by budget, and will be around 100–200 teraflops. Here, however, we will show that a petaflops machine is technologically feasible.

The main goal of the machine is to perform the simulation of collisional systems. We first describe what is possible: We can make a chip with a speed exceeding 100 Gflops using the technology of 1998. Then, we discuss several design choices for the basic design. Some of them will be discussed later in this chapter, in conjunction with the possibility of extending the functionality. We first discuss how the memory is attached to pipeline chips. Then we discuss the communication network and host interface.

7.3.1 Pipeline processor LSI

The basic design of GRAPE-6 will be optimized for a single problem, the long-term, high-accuracy integration of collisional N-body systems. Thus, the design of the force calculation pipeline is essentially the same as that for GRAPE-4, though we will make a few changes to optimize the design further. The relative accuracy of the force itself is that of IEEE-754 single precision, or slightly better but with a simpler round-off logic. The exponent will be extended to at least 11 bits. Positions are represented either in 64-bit floating-point or 80-bit fixed-point format, and the accumulation of the force is performed in 64-bit floating-point format. The accumulation might be done in 80- or 96-bit fixed-point format, to simplify the accumulators and communication network. In this case, however, we need to supply a scaling factor for the force on each particle. The calculation of the jerk is performed at somewhat lower accuracy. Velocities are given in single-precision format. In the predictor pipeline, higher order derivatives are given in decreasing mantissa lengths (shorter by four bits at each order).

The predictor pipeline will use the second time derivative of the acceleration in addition to the first derivative used in GRAPE-4. In most cases, it significantly improves the accuracy, and the additional cost of the hardware is very small.

The GRAPE-4 processor chip implemented the 1/3 version of the HARP pipeline, which requires three cycles to calculate one force. The reason why we implemented the 1/3 version was simply that a full pipeline could not fit in a chip using 1 μm technology. The HARP chip has die size of 14mm $\times$ 14mm, which was about the limit the chip foundry could manufacture within our budget for the chip.

For the VLSI technology of GRAPE-6, we assume the following:

- Total number of usable gates is 4 M, and the transistor density for SRAM is twice as that of the logics.
- Operating voltage is either 2.5 or 1.8 V.
- Clock frequency is 150 MHz for the pipeline and 25 MHz for chip-to-chip communication (long distance on board lines).
- The number of I/O pins available per chip is 3–400.

In GRAPE-4, a single force calculation pipeline consists of about 400K transistors, or about 100K gates. The ingredients are the following: a 64-bit floating-point adder consumes about 8K gates, and one 32-bit floating-point multiplier consumes 6K, and one adder 3K. The number of gates for a floating-point multiplier is almost proportional to p^2, where p is the square of the number of bits in mantissa, while that for the adder is roughly $p \log p$. In GRAPE-4, we used table lookup with quadratic interpolation to calculate $1/r$, etc. The table has 256 entries, each with a 50 bit width, to supply zeroth, first and second derivatives, which required 15K transistors for ROM. The hardcoded ROM requires just one transistor per bit, and the size of the transistor is significantly smaller than that used for the logic. So the silicon used for this table is tiny, even though we used three of these tables to calculate $1/r$, $1/r^3$ and $1/r^5$ in three cycles.

The clock speed of the GRAPE-4 chip is currently 32 MHz. Most chips could operate perfectly at 50 MHz. Unfortunately, the clock speed of the total system is limited by the clock speed of the floating-point adder chip used to collect the results, even though this part operates at a clock frequency half that of the pipeline chip. The listed speed of this adder chip is 25 MHz, but we were not able to make this chip reliably work at that clock speed.

If we had realized that the pipeline chip could actually work at 50 MHz, we would have increased the number of virtual pipelines with a possibly adjustable virtual/physical ratio, to optimize the ratio between the calculation speed and the communication speed.

The number of transistors available on the GRAPE-6 chip is about 40 times that of the GRAPE-4 chip, because of the reduction in the feature size and advances in the wiring technology. The feature size is reduced by a factor of four, resulting in an increase of the raw transistor count by a factor of 16. The additional increase of about a factor of two is achieved by increasing the number of wiring (metal) layers from two to five, and also by making the physical chip size a bit larger. Whether we can actually make a physically bigger chip within our budget will soon become clear.

A basic pipeline design would require about 250K gates or 1M transistors, if we design a pipeline which can deliver one result per cycle. Thus, we can pack about 16 pipelines onto a chip. In other words, we can integrate the present GRAPE-4 processor board with 48 pipeline chips (each requires three cycles to deliver one result) into a single chip.

The operating frequency of 150 MHz will not be too difficult, simply because the feature size is reduced by a factor of four. As stated, the present GRAPE-4 chip can operate at 50 MHz. The total speed would then be $0.15 \times 16 \times 60 = 144$ Gflops.

In the following, we will see whether we can utilize such a monstrous chip efficiently. We need to evaluate two factors: the first is the power consumption of a chip; the second is the required communication bandwidth. Since the communication bandwidth depends upon many factors, we shall analyze it in conjunction with the overall architecture in the next two sections. In this section, we give a brief estimate for the overall power consumption and requirement for cooling.

In the case of the GRAPE processor chip, the power consumption is almost entirely due to the transistors in the pipeline, since the chip-to-chip connection does not scale with the computation. To get some idea, let us use the GRAPE-4 chip as the baseline and extrapolate it to the GRAPE-6 chip. The GRAPE-4 chip dissipates about 5 W when operating at 32 MHz. The GRAPE-6 chip, with 200 times more performance, would dissipate 1KW if designed with the same technology. However, there are two factors which greatly reduce the power consumption. First, the shrink from 1 to 0.25 μm reduces the power consumption by a factor of 4. Then, the reduction of the power supply voltage from 5 V to perhaps 1.8 V reduces the power consumption by another factor of 8, resulting in a total reduction of a factor of 32, or about 30 W per chip. This is comparable to the power consumption of typical present-day microprocessors.

A petaflops system will thus generate about 250 KW, which is comparable to that of a present day large MPPs or vector-parallel computers.

The heat dissipation of 30 W per chip is, of course, quite large. If we use forced air cooling, it is pretty hard to remove more than 10KW of heat from a standard 19-inch rack. So we need around 30 racks, each with 300 chips, with perhaps about 20 chips per board. GRAPE-4 dissipates a maximum of 3KW per rack, which is quite easy to handle without serious cooling consideration.

7.3.2   Memory organization

A petaflops GRAPE-6 system will consist of 10 000 pipeline chips or about 160 000 force calculation pipelines. If we want to use virtual multiple pipelines to keep the external clock speed low, the total number of virtual pipelines would be around 1 million. How we organize this huge number of pipelines? In the case of GRAPE-4, the total number of virtual pipelines is less than 4000, and we calculate the forces on 96 particles in·parallel. Thus, the force on one particle is calculated as 36 partial sums on 36 chips. These the partial sums are first accumulated on the floating-point adder chips on the control boards, and then on the host computer.

For GRAPE-6, the number of particles on which the forces are calculated in parallel should not be much larger than the 96 of GRAPE-4, since the typical number of particles for which we use GRAPE-6 is less than 10^6. Even if we put this number at 1000, we need 1000 separate memory modules, instead of the 40 modules of GRAPE-4. In addition, each memory module needs to have a data width of 600 bits, instead of the 200 bits of GRAPE-4 processor boards. Thus, the connections between processor chips and memory chips in

GRAPE-6 need 75 times more wires. This is far too large, unless some exotic packaging technology is used.

In the following, we consider two alternative approaches. The first is to integrate the memory and processor pipelines into a single chip. The second is to connect them with some unusually fast connection.

First we consider the processor-memory integration. The total amount of memory needed for GRAPE-6 will be quite small. For any conceivable application, memory for 10^8 particles will be far more than enough, since a direct calculation of force for that many particles would be too costly. One timestep would take 10 minutes.

A petaflops system will consist of 10 000 chips. Therefore, the storage of 10K particles per chip is more than enough. The memory needed to store 10K particles is 5 Mbits, which is easy if DRAM can be integrated, and not impossible with SRAM. Already in 1995, the AMD SHARC (21060) chip from Analog Devices integrated 4 Mbits of SRAM with a DSP core.

The integration of the memory and processor has several important advantages. First, it improves the memory bandwidth drastically. The on-chip memory can operate at the clock speed of the chip, and the width of the data path to the pipeline processors is practically unlimited. Secondly, it reduces the I/O pin count and total number of wires on the board. Each processor chip has just one communication port, so the manufacturing costs of the board would become considerably smaller. In addition, we might be able to use inexpensive packages for the processor chip.

On the other hand, if we use 50% of the total real estate of the chip for the memory, we need to use twice as many chips to achieve the same overall performance. Thus, we need to do a fairly detailed cost estimate to determine whether this approach is really advantageous compared to a more conventional approach of adding external memory.

In the case of GRAPE-4, we let 48 chips share one memory unit. With GRAPE-6, however, such an option would limit the available memory bandwidth too much. Currently, both SRAM and DRAM memory chips can operate at clock frequencies exceeding 100 MHz. On the other hand, we plan to design the overall system so that it operates at a clock frequency of around 25 MHz. To design a board with a clock speed higher than 25 MHz is not easy, at least for us. Thus, we would waste the bandwidth of the memory chip if we let them operate at the board clock.

If we can keep the physical distance between the memory chip and the processor chip small, it would not be too difficult to let them operate at a much faster clock speed. Thus, one possibility is to attach a single fast memory chip to each processor chip. By using either the RAMBUS protocol or fast synchronous DRAM, we can achieve a data transfer speed of 500–600 MB/s. The high clock frequency is not too problematic, since the connection is limited to two neighboring chips. Since the data for one particle is around 500 bits, this bandwidth gives a data transfer speed of 10^7 particles/sec. On the other hand, the speed of one chip is around 2×10^9 interactions/sec. Thus, we should design a chip so that it calculates the force on 200 particles using both physical and virtual multiple pipelines. The number of physical pipelines will be 16: the number of virtual pipelines per physical pipeline will be therefore around 16.

7.3.3   Comparison of on- and off-chip memory

If we integrate the memory and the processor, extremely high bandwidth is not too difficult. In particular, with SRAM it would not be much of a problem to assign separate memory blocks to individual pipelines, so that all the pipelines in one chip calculate the force on just one particle. In particular, we could organize the whole system into a single reduction tree in such a way that all pipelines calculate the force on a single particle.

This monolithic single tree, however, does not work in practice, simply because a single chip cannot provide the necessary bandwidth of several GB/s. A petaflops system can calculate 1.6×10^{13} interactions per second. Thus, if we require the system to have acceptable performance for $N = 10^6$, the communication time to transfer the necessary data for one particle should be around 60 ns. The amount of data to be transferred for one particle is around 1000 bits. Thus, the necessary data transfer rate is 16 Gbits/s, or 2 GB/s.

If we construct a single tree, we need this speed of 2 GB/s at each link of the tree. If we are to use a 25 MHz clock for chip-to-chip communication, we need 640 pins to achieve the necessary communication speed. In practice, we would like to limit the number of I/O pins to 64, or a maximum of 128.

It is necessary that forces on several particles (around 20) are calculated simultaneously, just to be able to read them out in parallel. On the other hand, if we attach external memory, each chip calculates the force on around 200 particles.

An acceptable design is determined from the following consideration. Let c be the communication bandwidth of the processor chip for register read/write, and C the total bandwidth to the host. Then obviously, we need $n = C/c$ independent ports to host. This means that n particles are calculated concurrently, even if each chip calculates the force on only one particle.

The total number of particles on which the forces are calculated in parallel is then given by

$$f = np, \tag{7.1}$$

where p is the number of logical pipelines per chip, which is then expressed as

$$p = Pb_p/b_m, \tag{7.2}$$

where P is the physical number of pipelines per chip, b_p is the input data bandwidth of one pipeline, and b_m is the memory bandwidth per chip. We assume $C = 2\text{GB/s}$, $c = 0.2\text{GB/s}$, and therefore $n = 10$. For other parameters, we assume $P = 16$, and $b_p = 10\text{GB/s}$. In the case of the internal memory, we can set b_m to be close to Pb_p, and therefore p close to unity. Thus, $f < 100$ is not very difficult. This make the system relatively easy to use.

On the other hand, for the external memory, b_m exceeding 1GB/s is hard to implement. Therefore, we end up with $p = 160$, or $f = 1600$. This is quite large, though not really unacceptable.

The actual decision will depend upon more detailed analysis, but it seems obvious that the integration of memory and pipelines offers many advantages. In practice, the ease of manufacturing and testing the board will be the single most important issue.

7.3.4   Network structure

Here we consider two alternatives for the network which connects the processor chips and the host computer.

We do need some kind of tree structure to achieve the necessary bandwidth. In the case of GRAPE-4, we could have a hierarchy of bus-based connection, because the communication speed of the pipeline chip exceeded our requirements. However, with GRAPE-6, the communication speed of the lowest level (the processor chip) will fall far short of our requirements, and we will need parallelism to achieve the speed necessary. There is no room for bus-based connection, so a tree structure based on point-to-point connections is necessary.

There are probably two or three issues to be addressed, namely:

(a) Shall we use a binary tree or a tree with a higher radix?
(b) Shall we use the tree or shall we consider sorting networks, such as the Omega network?
(c) What physical connection shall we use?

First let us discuss the network topology, since it will have some impact on other issues. The tree radix has quite a small effect on the total overhead, but has a large impact on the number of parts and wires, in particular if an elaborate structure is used. Figure 96 shows a possible structure, for the case of a four-cluster, 16-processor-board system. Trees in thick lines are used to retrieve the result. The dedicated write channels in thin lines are used to broadcast the data to be written to the memory to all clusters.

This rather elaborate network allows us to access the chip registers in four clusters simultaneously, and to write memory data to all memory units, again simultaneously. This is achieved by having two diagonal tree structures, one within the cluster and the other over multiple clusters.

This structure is perfectly alright for dedicated use of the machine for a single problem, but it is difficult to change the configuration. In particular,

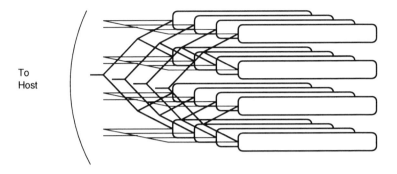

**To
Host**

Figure 96   A monolithic network configuration shown for four clusters, each
with four processor boards.

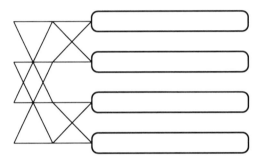

Figure 97   4 × 4 butterfly network.

there is no way to use the machine in a partitioned fashion, even though the
machine has multiple parallel interfaces to the front-end.

A natural way to implement a reconfigurable and reasonably scalable network is to use a butterfly network, which is shown in Figure 97. On paper, this
network looks perfect. First of all, it can replace the structure of the original
network quite nicely. It can perform all three necessary forms of operation:
multiple broadcast to the memories, multiple broadcast to the registers, and
multiple summation and retrieval from the chip registers. More importantly,
this network can be partitioned into subnetworks of smaller sizes.

One practical problem with the butterfly network is that it requires too
much hardware. For the total system of 256 boards, the original configuration
requires only 240 tree nodes and 16 broadcast units, and the total number
of connections is similar. Here, the largest concern is in the total number of
board-to-board links, which is about 500.

The full butterfly network bursts the number of nodes from around 500 to around 2000. A practical concern here is that a butterfly switch, unlike the single binary tree, requires long wires between switches. A long wire is a guaranteed way to create trouble in a high-speed digital circuit, and particularly so if we try to push the clock speed to more than 25 MHz and the bus width to 64 bits or more. In the design of GRAPE-4, we were quite careful to keep the length of the bus line driven by an FPGA chip to be as short as possible. This principle is, however, rather difficult to maintain.

An attractive possibility is to use a fast serial connection for the link between network switches. A serial connection can achieve a speed of 1 Gbits/s without much difficulty.

To summarize, we believe the butterfly network is far easier to use than the monolithic configuration previously described. The radix of the tree is more of a matter for the implementation details.

7.3.5   An overview

Figure 98 gives an overview of an example of the petaflops GRAPE-6 system. It consist of 16 384 processor chips, each containing 3 Mbits of SRAM memory and eight pipeline processors. One processor board consists of 32 pipeline chips, and one cluster consists of 16 processor boards. The total system consists of 32 processor clusters and one control box. The host and control box are connected with 32 serial links, each with a bandwidth of 100 MB/s. The control box implements 32 trees, so that the data from all host links can be broadcast to all processor boxes simultaneously. Each processor box is connected to the control box again by 32 serial links.

Each processor box contains 512 processor chips, which are organized in a 32×32 butterfly network on top of a four-level binary tree. This system will be able to deliver an effective speed exceeding 500 Tflops for the simulation of globular clusters with 500 000 stars.

The development cost of the system is dominated by the cost of developing the pipeline chip, which will be around $ 1 million. The production cost of a petaflops system with 16 384 chips will be about 10 times that of the production cost of the 1700-chip GRAPE-4 system. Thus, the total cost of the petaflops system will be around $ 10 million.

For a smaller grant, like $ 2-3 million for hardware development, the practical speed goal will be around 1–200 Tflops. In this case, the overall design could be quite different. The number of pipeline chips will be around 2–4000, in other words, similar to that of GRAPE-4. On the other hand, the requirement for the total amount of memory is not much different, since the number of particles that a 100 Tflops system can handle is only a factor of two less than that which can be handled with a 1 Pflops system. The requirement for communication, on the other hand, is reduced by a factor of four, going down to around 500 MB/s. Thus, for a speed in the range of 100–200 Tflops, the design with an external memory would be more practical.

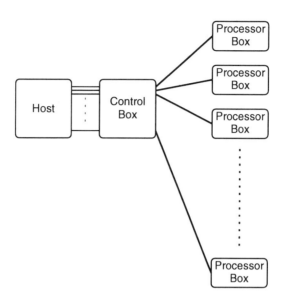

Figure 98   The GRAPE-6 system.

7.4   Widening the application area

In this section, we consider several possibilities to widen the application area
of the GRAPE system, using various approaches.

7.4.1   Arbitrary force law

The inclusion of the arbitrary force law would increase the potential application of GRAPE-6 significantly. One question here, however, is what level of
generality will we have to implement?

There are three main application areas for which the programmable force
law is important. The first is the calculation of gravity or Coulomb force
under a periodic boundary. We need a truncation for the Coulomb force.
Both the Ewald method and the P^3M method require the direct force to be
truncated, since the potential the field of the long wavelength is taken care by
the Fourier transform. This Coulomb force with truncation must be calculated
for a wide range of the distance, but for most of the range the deviation from
the untruncated force is negligible.

The second area is the van der Waals potential, which goes off as r^{-6}. This
force must be defined for the dynamic range of something like $1/2$ of the
inter-atom distance to 5–10 times that, with varying degrees of accuracy.

The third area is the calculation of SPH interaction. This will be discussed later, since SPH interaction is not a simple central force and requires more complex hardware.

The inclusion of the programmable force table, by itself, is not very hard. Depending on the level of accuracy, we need around 100 Kbits for a lookup table and around 10K gates for the interpolator. Depending on the design method available, these can consume 20–40% of the real estate of the pipeline chip, thus resulting in the same amount of reduction in the total performance.

However, to implement the short-range van der Waals force is not just a matter of the lookup table. In the typical case, the van der Waals force between atoms i and j depends upon the *pair* of the types of the atoms. The atom type then depends upon the location of that particular atom in the molecule, and the number of types can be as large as 100. Each atom pair is expressed by two coefficients, resulting in another 640 Kbits of table for each pipeline.

The table size of 640 Kbits per pipeline is far too large to be considered seriously. This table is not present in the MD-GRAPE chip, which made the chip quite difficult to use in real applications. A MD-GRAPE board has only one memory for this table. Particles on which the forces are calculated in parallel must be of the same type.

With GRAPE-6, one possibility is to let multiple pipelines share the same table, and the other possibility is to allow the chip to load the table from off-chip memory before calculation. Yet another possibility is to have one big memory shared by pipelines, with cache memories for individual pipelines. Note that the "hit-rate" of this cache would be very high, because the type of atom on which the force is calculated is unchanged during calculation. However, it would complicate the design further, resulting in a longer development time.

To make the chip design still more complicated, the atoms in the same molecular chain do not directly interact through the van der Waals force. So it is highly desirable that there is some easy way to mask specified interactions, which can be accomplished with the bitmask table. This table is a small fraction of the table for the particle data itself, and therefore is not very costly. This table can also be used to reduce the communication in the gravitational simulation with an individual timestep.

The important decision here is whether to implement the van der Waals force in hardware, because the number of atoms interacting through the van der Waals force is very limited. On the other hand, the calculation cost of the Coulomb force is much higher. Even with the Ewald method, for a large number of particles the cost of Coulomb calculation can be more than 99% of the total calculation cost, which implies that we can accelerate the calculation by a factor of 100 by putting the Coulomb force in hardware, without implementing the capability of handling the van der Waals force in hardware.

Thus, one quite viable possibility is to add the functionality to perform the Ewald (or P^3M) algorithm in hardware, but not to add the functionality to calculate the van der Waals force.

If we limit the main application to the Ewald/P^3M method, the implementation is quite significantly simplified, because the only thing we need to implement now is the cutoff, which can be implemented as a separate, relatively small table of a limited dynamic range.

7.4.2 SPH

Whether to implement SPH is yet another question. The main use of GRAPE outside our group is for SPH (see Section 6.5.2). Here, we briefly summarize what functionalities are necessary to implement SPH in hardware, and what the cost of that would be.

To implement SPH in hardware, we need to implement several different types of operation, namely

$$< f_i > \quad = \quad \sum_j f_j W(r_{ij}, h_{ij}), \tag{7.3}$$

$$< \mathbf{g}_i > \quad = \quad \sum_j f_j \nabla W(r_{ij}, h_{ij}), \tag{7.4}$$

$$< f_i > \quad = \quad \sum_j \mathbf{g}_j \cdot \nabla W(r_{ij}, h_{ij}). \tag{7.5}$$

As far as we use a spherical kernel, we can use the relation

$$\nabla W = \frac{dW}{dr} \frac{\mathbf{r}}{r} \tag{7.6}$$

to evaluate the latter two formulae.

To make the matter more complicated, there are several variants for the way in which the kernel function W is symmetrized, and there are also several different ways to handle the artificial viscosity, even if we neglect exotic algorithms such as ASPH [SMVO96]. The usual way in which the artificial viscosity is implemented is through the Monaghan–Gingold tensor [GM77]:

$$Q_{ij} \quad = \quad \begin{cases} \frac{-\alpha c_{ij}\mu_{ij}+\beta\mu_{ij}^2}{\rho_{ij}}, & (\mathbf{r}_i - \mathbf{r}_j) \cdot (\mathbf{v}_i - \mathbf{v}_j) < 0 \\ 0, & \text{otherwise,} \end{cases} \tag{7.7}$$

$$\mu_{ij} \quad = \quad \frac{h_{ij}(\mathbf{r}_i - \mathbf{r}_j) \cdot (\mathbf{v}_i - \mathbf{v}_j)}{r_{ij}^2 + \eta^2}. \tag{7.8}$$

This "standard" viscosity, however, is known to be ill-behaved in the case of the shear flow, and several modifications have been proposed (see, for example, Balsara [Bal95]).

To implement hardware which can accommodate more than one implementation of SPH is difficult, and it might become useless if a better algorithm is found.

One practical possibility is to use an FPGA (Field Programmable Gate Array) to implement the SPH calculation. Buell *et al.* [BA96] described the use of FPGA to implement various computational tasks on their SPLASH II hardware. Whether or not this is feasible depends very much upon the required relative accuracy of the SPH interaction. Preliminary studies [Mas95], [YOT+96] suggest that the required relative accuracy is actually rather low and an error larger than 1% is acceptable. In this case, it is feasible to implement one full SPH pipeline into a single FPGA chip of today, and multiple pipelines can be implemented with larger chips, though the advance in the density of FPGA devices is not as fast as that of custom VLSI chips.

For SPH particles, we are not much concerned with the maximum number of parallel force calculations, since the number of particles with the smallest timestep for SPH calculation is likely to be fairly large. In this case, many SPH pipelines can share one big memory unit, in much the same way as in GRAPE-3/4 systems. Since the memory for the gravitational force calculation is integrated in the pipeline chip, this implies that the memory for the SPH pipelines will be separate.

This separate structure, however, poses a problem. How can we create and use the neighbor list efficiently? Each GRAPE pipeline can construct the neighbor list for one particle, which is stored in the memory within the chip. Unless each SPH pipeline has its own memory unit, however, we cannot use it. If we have multiple pipelines which share one memory unit, all of them naturally calculate the contributions from the same set of particles. There is no simple way to circumvent this problem.

GRAPE-like machines for molecular dynamics calculation had faced the same problem, since the van der Waals force has a much shorter effective range than that of the Coulomb force, even when the latter is truncated at a finite distance. Two machines built recently (MD-GRAPE and MD-Engine) used quite different approaches. In the case of MD-GRAPE, the designers basically gave up using the neighbor list, and adopted the usual linked-list approach [HE88]. The memory and address generator is designed so that the linked-list method can be run efficiently, but this did cause a fairly large increase in the total calculation cost.

In the case of the MD-Engine, each pipeline is equipped with its own memory, so that it can actually use the neighbor list.

If we go back to older machines developed before GRAPE, the Delft Molecular Dynamics Processor [BB88] used the linked list, while FASTRUN [FDL91] used the neighbor list.

In the case of the hardware for molecular dynamics calculation, the strategy of DMDP/MD-GRAPE was clearly better than that of FASTRUN/MD-Engine, since this choice resulted in nearly a factor of 10 difference in the raw performance of the pipeline. The maximum gain one can get from the neighbor list is a factor of only a few. So the simpler design of MD-GRAPE is more attractive.

However, for SPH hardware the decision is more difficult. If we are to use FPGA chips to implement SPH functions, the raw speed of the SPH functions will be quite significantly slower than the raw speed of the gravitational force pipelines, because of the low degree of integration. This by itself is alright, since the calculation cost of SPH is much smaller than that of gravity by at least a factor of 10, and possibly 50 or more. However, if we cannot utilize the neighbor list, this ratio becomes much closer to unity.

In practice, another factor is that in many applications only a small fraction of particles are SPH particles, and the majority of particles in the system are collisionless. So the use of the linked list, or some variation of it, might not actually be too bad.

If we do not use the neighbor list for SPH, the simplest implementation of the SPH function is to have separate processor boards dedicated to SPH calculation. It would consist of a large memory and many FPGA chips. In fact, such a system might become commercially available from companies like Annapolis Micro Systems, which sells a reconfigurable computing engine based on Xilinx FPGA. We should evaluate these systems to see if they can deliver sufficient performance to match the speed of the petaflops GRAPE.

In many cosmological calculations, the number of SPH particles is less than 50% of the total number of particles, and in many cases around 10%. This factor, combined with the difference in the number of interactions, should result in at least a factor of 100 difference in the total calculation cost, even when the tree algorithm is used for gravitational force calculation. Thus, the SPH system with a speed of around 1Tflops would be sufficiently fast.

7.4.3   Non-direct force calculation

Here we evaluate the gain we can achieve by combining GRAPE with sophisticated force-calculation algorithms such as Barnes–Hut treecode and FMM. The gain for BH treecode is analyzed in detail by Makino [Mak91c]. The bottom line is that the calculation time will be dominated by the time taken to construct the tree and to calculate the multipole moments for the tree nodes, if GRAPE is sufficiently fast. These typically represent 1–2% of the total CPU time on a reasonably balanced single-processor RISC workstation, and therefore one can get up to 50 times the speed of the calculation on the front-end.

The same is essentially true for FMM. The difference here is that with FMM, high-level node–node interaction is handled on the front-end. GRAPE handles the particle–particle interaction at the lowest level of the tree.

Roughly speaking, the calculation cost of an FMM scheme per timestep is expressed as

$$C \sim c_1 N n_p + c_2 N p^2 + 200 c_3 n_n p^2 \log p, \qquad (7.9)$$

where c_1, c_2 and c_3 are constants which depend upon the speed of the computer, N is the total number of particles, n_p is the average number of particles in a node at the lowest level of the tree, p is the degree of multipole expansion, and n_n is the total number of nodes. Naturally, we have $N = n_p n_n$. The constant 200 is the typical number of nodes with which a node interacts, which depends to an extent on the implementation and the accuracy required. Here we assume that we use a highly sophisticated scheme which involves the use of FFT to shift polynomials [BHER95]. Actual timing comparisons have shown that an FFT-based scheme is marginally faster than a straightforward implementation which scales as p^4.

The total cost C is the minimum for

$$n_n = \frac{N}{p} \sqrt{\frac{c_1}{200 c_3 \log p}}, \qquad (7.10)$$

and the minimum value is

$$C_{min} = N p \sqrt{800 c_1 c_3 \log p} + c_2 N p^2. \qquad (7.11)$$

In practice, it is possible to reduce the second term by having the hierarchical structure within the lowest level. It can be shown that if n_p is large enough, the second term becomes independent of p.

If the first term dominates, the above equation implies that the total calculation cost is proportional to $\sqrt{c_1 c_3}$. In other words, the overall speed is the geometric mean of the speed of GRAPE and that of the front-end. Thus, we can achieve about 30 times better performance if GRAPE is 1000 times faster than the front-end. For the same cost, GRAPE will be about 1000 times faster than a general-purpose computer, and therefore it will accelerate the FMM scheme by a factor of 30.

One could imagine hardware which performs polynomial shifting, or simply FFT calculation for small data, to further augment the speed of FMM. Whether or not such hardware is practical requires further study. Here, again, the use of an FPGA-based device could be helpful. The accuracy required for the hardware is determined by the accuracy required by the application, and the possibility of customizing the hardware for the required accuracy might offer quite a high performance for certain applications.

7.5 GRAPE entirely based on FPGA?

We have seen that reconfigurable computing engines based on FPGA might be well suited for various needs, such as implementing SPH calculation or an FMM scheme. Well, then, why not proceed one step further to implement the entire GRAPE hardware itself using FPGA?

The reason not to do so is that the performance which can be achieved with FPGA is at least two orders of magnitude lower than that which can be achieved by custom LSIs. Using the $0.25\mu m$ technology available now, one can pack about 4 million gates in a custom LSI. On the other hand, the largest FPGA has a nominal gate count of around 150K, but the actual size of the circuit that can fit into that particular chip is typically less than 40K gates. Thus, there is about a factor of 100 difference in the density. In addition, there is a factor of a few difference in the clock frequency. These two factors have been roughly constant from the time when FPGA was first introduced. It is not likely to become smaller in the predictable future.

If we use the FPGA as the building blocks for GRAPE-6, the price-performance would go down by this factor of 100, and a petaflops system would cost nearly $ 1 billion. Thus, to build a big machine entirely based on FPGA is not a good strategy.

In the case of SPH calculation of self-gravitating systems, it makes sense to use FPGA to handle SPH interaction, simply because the SPH part is not as costly as the gravitational interaction. We can accept rather a low cost-performance of an FPGA implementation for SPH, for the same reason as we can use slow general-purpose computers to handle time integration of the orbit and other tasks.

To summarize, a "traditional" GRAPE system is a heterogeneous system which consists of GRAPE hardware to handle direct calculation of gravitational force and a front-end to handle all the other calculations. This architecture is fine for gravitational N-body simulation, but the gain in performance was rather limited in the case of SPH calculation or sophisticated algorithms, because the total performance is limited by that of the front-end.

To add the third piece of hardware, namely the reconfigurable computing engine, to handle moderately compute-intensive tasks such as evaluation of the SPH interaction, could greatly improve the total performance of the system for wide range of applications.

7.6   Applicability of special-purpose devices for applications other than N-body simulations

One of the reasons for the success of the GRAPE system is clearly the nature of the problem. The target application is the gravitational N-body problem, where N particles in the system interact with all other $N-1$ particles. In addition, the requirement for relatively high accuracy and a high degree of spatial inhomogeneity makes it impractical to use fast and approximate schemes such as particle-mesh or Barnes–Hut treecode.

In other words, the success of GRAPE is largely due to the rather unique nature of the application. In the following, we will briefly discuss whether special-purpose computers like GRAPE will have a wider range of application or not.

7.6.1   Systems with short-range interaction

As discussed in Chapter 3, a number of special-purpose computers have been built for systems with short-range interactions, such as the Ising model and Lattice QCD. As we have seen so far, they have been quite successful. Ising Spin processors used a highly specialized processor, while QCD machines used programmable, off-the-shelf processors as the building block.

The viability of the latter approach is somewhat questionable, since commercial parallel computers now rely on the same approach. As long as the same microprocessor is used, there is not much reason to expect a large gain in price-performance. If special-purpose QCD machines are to achieve a much better price-performance than commercial parallel computers, they need to use something other than existing microprocessors. However, to develop a programmable parallel computer without using an existing microprocessor (and the software developed for it) is a highly costly venture, with the development cost being well over $ 10 million. This rather high development cost requires that the machine can be used for many different applications, not just for a single problem of (for example) calculating the proton mass. The need for the generality has a negative impact on the performance.

The approach of building specialized processor is perhaps more promising. The reason why Ising machines lost the advantage in performance is that their performance is limited by the memory bandwidth. The integration of the memory and processor can provide an enormous advantage of practically unlimited memory bandwidth.

Even with the integrated memory, the need to communicate with the other processors would eventually limit the performance of a single chip. Let us illustrate this by a paper design for a special-purpose computer for three-dimensional computational fluid dynamics on a regular grid.

If a chip comprises the memory for M grid point and p arithmetic units, it takes

$$n = CM/p \qquad (7.12)$$

cycles per one timestep, where C is the number of floating point operations per grid point per timestep. Thus, if we keep the ratio M/p constant, n is constant. To obtain the data for the boundary, the chip must send/receive the data for $6M^{2/3}$ grid points in n cycles. If we keep n constant, the number of connections increases as $M^{2/3}$. In other words, the number of pins should increase as 2/3 power of the total number of transistors on the chip, even if the I/O pins operate at the same clock speed. This assumption is clearly far too optimistic. Thus, asymptotically, we need a very large number of pins. In other words, to make a practical design, we have to reduce the ratio p/M. If this ratio becomes too small, the relative advantage over a general-purpose processor-in-memory architecture will vanish, in the same way as the advantage of Ising processors had vanished.

Whether such a diminishing return will actually cause a problem is, however, a different question, which must be evaluated with real numbers based on the available technology. With 0.25 μm technology, we can integrate 3×10^7 bits of memory and, in theory, about 100 64-bit floating-point units (total of adders and multipliers), if we give equal areas to processors and memory. The minimal amount of memory per one grid point is around 300 bits, and the number of floating-point operations per grid point for complex, high-order schemes is around 500. Thus, we have $p = 100$, $M = 10^5$ and $C = 500$, and therefore $n = 5 \times 10^5$. The number of bits to be communicated is $12 \times 300M^{2/3} \sim 7 \times 10^6$. Here the factor 12 comes from the six sides of a box and the need for bidirectional transfer. The amount of data transfer per cycle is around 15 bits. This looks like a small number, but we have to transfer this amount of data for each clock cycle. On the other hand, the clock frequency for the external connection is currently much lower than that of the internal clock speed. Thus, the actual number of pins required is several times larger than this 15, and it will increase very rapidly. Thus, surprisingly, the performance of a special-purpose computer for short-range interaction is already limited by the speed of the chip-to-chip communication bandwidth, and therefore it does not have an advantage over the general-purpose PIM architecture.

In reality, the relative advantage of a special-purpose computer depends upon the ratio of the memory and processing power of general-purpose PIMs. Thus, even though the theoretical advantage is not too large, the actual advantage over commercially available machines can be much larger.

7.6.2 Systems with long-range interaction

For systems with long-range interaction, the GRAPE-type approach is certainly better than a general-purpose parallel computer. For the direct summation algorithm, the relative advantage is not likely to diminish.

However, whether the direct summation is a good solution for a very large number of particles is a different question. It is unlikely that complex algorithms such as parallel FMM [BHER95] can give a large performance gain for the number of particles less than 10^6. However, for larger simulations, these schemes performed on a programmable machine with carefully designed communication network might outperform direct summation, even when high accuracy is required.

This is clearly an area which needs further research, in particular in the field of the algorithm for inhomogeneous systems.

Algorithms like Barnes–Hut tree and FMM can also be accelerated with a GRAPE-type machine. However, the gain will ultimately be limited by the data communication bandwidth between the special-purpose part and the general-purpose part, if we retain the same architecture as we are using today. In other words, eventually it will be necessary to integrate the general- and special-purpose parts more closely if we want to achieve a high performance for sophisticated algorithms.

7.7 Summary

In this chapter, we have overviewed the future of special-purpose computing. As an example, we used GRAPE-6, which will be completed by 2000–2001 with a peak speed of 100–200 Tflops. For the problems for which GRAPE-6 is designed, it will offer performance three orders of magnitude better than what will be available on general-purpose computers of a similar price tag.

Whether special-purpose computing will be a viable alternative to general-purpose computing in the future is, however, difficult to predict. If general-purpose machines evolve following the present trend, the relative advantage of a GRAPE-type machine will continue to increase. However, as in the case of QCD machines, a drastic change in the architecture of general-purpose machines can, in theory, eliminate the relative advantage of the GRAPE-type architecture. Whether such a change will occur is an open question.

References

[Aar63] Aarseth Sverre J. (1963) Dynamical evolution of clusters of galaxies, i. *Monthly Notices of Royal Astronomical Society* 126: 223–255.

[Aar85] Aarseth S. J. (1985) Direct methods for n-body simulations. In Blackbill J. U. and Cohen B. I. (eds) *Multiple Time Scales*, pages 377–418. Academic Press, New York.

[Aar96] Aarseth S. J. (1996) Star cluster simulations on harp. In Hut and Makino [HM96], pages 161–170.

[AC73] Ahmad A. and Cohen L. (1973) A numerical integration scheme for the n-body gravitational problem. *Journal of Computational Physics* 12: 389–402.

[AHW74] Aarseth S. J., Henon M., and Wielen R. (1974) A comparison of numerical methods for the study of star cluster dynamics. *Astronomy and Astrophysics* 37: 183–187.

[Ant62] Antonov V. A. (1962) Most provable phase distribution in spherical star systems and conditions for its existence. *Vest. Leningrad Univ.* 7: 135.

[APB97a] Athanassoula E., Puerari I., and Bosma A. (1997) Evolution of compact groups of galaxies i. merging rates. *Monthly Notices of Royal Astronomical Society* 286: 284–302.

[APB97b] Athanassoula E., Puerari I., and Bosma A. (1997) Formation of rings in galactic disks by infalling small companions. *Monthly Notices of Royal Astronomical Society* 286: 284–302.

[AW85] Aguilar L. A. and White S. D. M. (1985) Tidal interactions between spherical galaxies. *The Astrophysical Journal* 295: 374–387.

[AZ74] Aarseth S. J. and Zare K. (1974) A regularization of the three-body problem. *Celestial Mechanics* 10: 185–205.

[BA96] Buell D. and Arnold J.M.and Kleinfelder W. (1996) *Splash 2: FPGAs in a Custom Computing Machine*. IEEE Comp. Soc. Press, Los Alamitos, CA.

[Bal95] Balsara Dinshaw S. (1995) von neumann stability analysis of smoothed particle hydrodynamics—suggestions for optimal algorithms. *Journal of Computational Physics* 121: 357–372.

[Bar90] Barnes J. E. (1990) A modified tree code: don't laugh; it runs. *Journal of Computational Physics* 87: 161–170.

[BB88] Bakker A. F. and Bruin C. (1988) Design and implementaion of the delft molecular-dynamics processor. In Alder B. J. (ed) *Special Purpose Computers*, pages 183–232. Academic Press, San Diego.

[BBR80] Begelman M. C., Blandford R. D., and Rees M. J. (1980) Massive black hole binaries in active galactic nuclei. *Nature* 287: 307–309.

[Ber85] Bernacca P. L. (1985) Project hipparcos. *Astrophysics and Space Science* 110: 21–45.

[BGGT90] Bakker A. F., Gilmer G. H., Grabow M. H., and Thompson K. (1990) A special purpose computer for molecular dyanamics calculations. *Journal of Computational Physics* 90: 313–335.

[BGT⁺94] Banday A. J., Gorski K. M., Tenorio L., Wright E. L., Smoot G. F., Lineweaver C. H., Kogut A., Hinshaw G., and Bennett C. L. (1994) On the rms anisotropy at 7 deg and 10 deg observed in the cobe-dmr two year sky maps. *The Astrophysical Journal Letters* 436: L99–L102.

[BH86] Barnes J. and Hut P. (1986) A hiearchical o(nlogn) force calculation algorithm. *Nature* 324: 446–449.

[BH92] Barnes J. E. and Hernquist L. (1992) Dynamics of interacting galaxies. *Annual Review of Astronomy and Astrophysics* 30: 705–742.

[BHER95] Board J. A. J., Hakura Z. S., Elliott W. D., and Rankin W. T. (1995) Scalable variants of multipole-based algorithms for molecular dynamics applications. In Bailey D. H., Bjorstad P. E., Gilbert J. R., Mascagni M. V., Schreiber R. S., Simon H. D., Torczon V. J., and Watson L. T. (eds) *Proceedings of the Seventh SIAM Conference on Parallel Processing for Scientific Computing*, pages 295–300. SIAM, Philadelphia.

[BS84] Bettwieser E. and Sugimoto D. (1984) Post-collapse evolution and gravothermal oscillation of globular clusters. *Monthly Notices of Royal Astronomical Society* 208: 493–509.

[BSO95] Brieu P. P., Summers F. J., and Ostriker J. P. (1995) Cosmological simulations using special purpose computers: Implementing p 3m on grape. *The Astrophysical Journal* 453: 566–573.

[BT87] Binney J. and Tremaine S. (1987) *Galactic Dynamics*. Princeton University Press, Princeton. This is a full BOOK entry.

[Car86] Carlberg R. G. (1986) The phase space density in elliptical galaxies. *The Astrophysical Journal* 310: 593–596.

[CH84] Cohn H. and Hut P. (1984) Is there life after core collapse in globular clusters? *The Astrophysical Journal Letters* 277: L45–L48.

[Cha43] Chandrasekhar S. (1943) *Principles of Stellar Dynamics*. Dover, New York.

[CHW89] Cohn H., Hut P., and Wise M. (1989) Gravothermal oscillations after core collapse in globular cluster evolution. *The Astrophysical Journal* 342: 814–822.

[CIH⁺87] Chikada Y., Ishiguro M., Hirabayashi H., Morimoto M., Morita K.-I., Kanzawa T., Iwashita H., Nakajima K., Ishikawa S.-I., Takahashi T., Handa K., Kasuga T., Okumura S., Miyazawa T., Nakazuru T., Miura K., and Nagasawa S. (1987) A 6 × 320 mhz 1024-channel fft cross-spectrum analyzer for radio astronomy. *Proceedings of IEEE* 75: 1203–1210.

[CO85] Condon J. H. and Ogielski T. (1985) Fast special purpose computer for monte carlo simulations in statistical physics. *Reviews of Scientific Instruments* 56: 1691–1696.

[Coh80] Cohn H. (1980) Late core collapse in star clusters and the gravothermal instability. *The Astrophysical Journal* 242: 765–771.

[CW90] Chernoff D. F. and Weinberg M. D. (1990) Evolution of globular clusters in the galaxy. *The Astrophysical Journal* 351: 121–156.

[DC91] Dubinski J. and Carlberg R. G. (1991) The structure of cold dark matter halos. *The Astrophysical Journal* 378: 496–503.

[DD87] Djorgovski S. and Davis M. (1987) Fundamental properties of elliptical galaxies. *The Astrophysical Journal* 313: 59–68.

[DK86] Djorgovski S. and King I. R. (1986) A preliminary survey of collapsed cores in globular clusters. *The Astrophysical Journal Letters* 305: L61–L65.

[EIN93] Emori H., Ida S., and Nakazawa K. (1993) New method for the numerical integration of an n-body system in an external potential. *Publications of the Astronomical Society of Japan* 45: 321–327.

[EMO91] Ebisuzaki T., Makino J., and Okumura S. K. (1991) Merging of two galaxies with central black holes. *Nature* 352: 212–214.

[Ewa21] Ewald P. P. (1921) Die berechnung optischer und electrostatischer gitterpotentiale. *Ann. Phys.* 64: 253–287.

[FDL91] Fine R., Dimmler G., and Levinthal C. (1991) Fastrun: A special purpose, hardwired computer for molecular simulation. *PROTEINS:Structure, Function, and Genetics* 11: 242–253.

[FEM92] Fukushige T., Ebisuzaki T., and Makino J. (1992) Rapid orbital decay of a black hole binary in merging galaxies. *Publications of the Astronomical Society of Japan* 44: 281–289.

[FH95] Fukushige T. and Heggie D. C. (1995) Pre-collapse evolution of galactic globular clusters. *Monthly Notices of Royal Astronomical Society* 276: 206–218.

[FIM$^+$91] Fukushige T., Ito T., Makino J., Ebisuzaki T., Sugimoto D., and Umemura M. (1991) Grape-1a: Special-purpose computer for n-body simulation with a tree code. *Publications of the Astronomical Society of Japan* 43: 841–858.

[FM97a] Fukushige T. and Makino J. (1997) On the origin of cusps in dark matter halos. *The Astrophysical Journal Letters* 477: L9–12.

[FM97b] Funato Y. and Makino J. (1997) Change in mass and energy through collisions of two identical galaxies. submitted to The Astrophysical Journal .

[FME92a] Funato Y., Makino J., and Ebisuzaki T. (1992) Energy segregation through violent relaxation. *Publications of the Astronomical Society of Japan* 44: 291–301.

[FME92b] Funato Y., Makino J., and Ebisuzaki T. (1992) Violent relaxation is not a relaxation process. *Publications of the Astronomical Society of Japan* 44: 613–621.

[FME93] Funato Y., Makino J., and Ebisuzaki T. (1993) Evolution of clusters of galaxies. *Publications of the Astronomical Society of Japan* 45: 289–302.

[FME94] Funato Y., Makino J., and Ebisuzaki T. (1994) Attenuation of luminosity of distant galaxies by multiple gravitational lensing. *The Astrophysical Journal Letters* 424: L17–L20.

[FMI$^+$93] Fukushige T., Makino J., Ito T., Okumura S. K., Ebisuzaki T., and Sugimoto D. (1993) Wine-1: Special-purpose computer forn-body simulations with a periodic boundary condition. *Publications of the Astronomical Society of Japan* 45: 361–375.

[FMNE95] Fukushige T., Makino J., Nishimura O., and Ebisuzaki T. (1995) Smoothing of the anisotropy of the cosmic background radiation by multiple gravitational scattering. *Publications of the Astronomical Society of Japan* 47: 493–508.

[Fox88] Fox G. C. (1988) The hypercube and the caltech concurrent computation program: A microcosm of parallel computing. In Alder B. J. (ed) *Special Purpose Computers*, pages 1–40. Academic Press, San Diego.

[FSD83] Farouki R. T., Shapiro S. L., and Duncan M. J. (1983) Hierarchical merging and the structure of elliptical galaxies. *The Astrophysical Journal* 265: 597–605.

[FTM$^+$96] Fukushige T., Taiji M., Makino J., Ebisuzaki T., and Sugimoto D. (1996) A highly-parallelized special-purchase computer for many-body simulations with an arbitrary central force: Md-grape. *The Astrophysical Journal* 468: 51–61.

[Fuk96] Fukushige T. (1996) Gravitational scattering experiments in infinite homogeneous n-body systems. In Muzzio J., Ferraz-Mello S., and Henrard J. (eds) *Chaos in Gravitational N-body systems*, pages 279–284. Kluwer, Dordrecht.

[FVF+94] Ferrarese L., Van Den Bosch F. C., Ford H. C., Jaffe W., and O'Connell R. W. (1994) Hubble space telescope photometry of the central regions of virgo cluster elliptical galaxies. 3: Brightness profiles. *The Astronomical Journal* 108: 1598–1609.

[FWM94] Fox G. C., Williams R. D., and Messina P. C. (1994) *Parallel Computing Works!* Morgan Kaufmann, San Francisco, California.

[GG79] Gunn J. E. and Griffin R. F. (1979) Dynamical studies of globular clusters based on photoelectric radial velocities of individual stars. i – m3. *The Astronomical Journal* 84: 752–773.

[GH89] Goodman J. and Hut P. (1989) Primordial binaries and globular cluster evolution. *Nature* 339: 40–42.

[GHH93] Goodman J., Heggie D. C., and Hut P. (1993) On the exponential instability of n-body systems. *The Astrophysical Journal* 415: 715+.

[GM77] Gingold R. A. and Monaghan J. J. (1977) Smoothed particle hydrodynamics – theory and application to non-spherical stars. *Monthly Notices of Royal Astronomical Society* 181: 375–389.

[GO97] Gnedin O. Y. and Ostriker J. P. (1997) Destruction of the galactic globular cluster system. *The Astrophysical Journal* 474: 223+.

[Goo84] Goodman J. (1984) Homologous evolution of stellar systems after core collapse. *The Astrophysical Journal* 280: 298–312.

[Goo87] Goodman J. (1987) On gravothermal oscillations. *The Astrophysical Journal* 313: 576–595.

[GRA+96] Gebhardt K., Richstone D., Ajhar E. A., Lauer T. R., Byun Y.-I., Kormendy J., Dressler A., Faber S. M., Grillmair C., and Tremaine S. (1996) The centers of early-type galaxies with hst. iii. non-parametric recovery of stellar luminosity distribution. *The Astronomical Journal* 112: 105+.

[GS86] Gurzadian V. G. and Savvidy G. K. (1986) Collective relaxation of stellar systems. *Astronomy and Astrophysics* 160: 203–210.

[HA92] Heggie D. C. and Aarseth S. J. (1992) Dynamical effects of primordial binaries in star clusters. i – equal masses. *Monthly Notices of Royal Astronomical Society* 257: 513–536.

[HCB88] Hoogland A., Compagner A., and Blöte H. W. J. (1988) The delft ising system processor. In Alder B. J. (ed) *Special Purpose Computers*, pages 233–280. Academic Press, San Diego.

[HD92] Hut P. and Djorgovski S. (1992) Rates of collapse and evaporation of globular clusters. *Nature* 359: 806–808.

[HDC93] Huang S., Dubinski J., and Carlberg R. G. (1993) Orbital deflections in n-body systems. *The Astrophysical Journal* 404: 73–80.

[HE88] Hockney R. W. and Eastwood J. W. (1988) *Computer Simulation Using Particles*. IOP Publishing, Ltd., Bristol.

[Heg74] Heggie D. C. (1974) A global regularisation of the gravitational n-body problem. *Celestial Mechanics* 10: 217–241.

[Heg75] Heggie D. C. (1975) Binary evolution in stellar dynamics. *Monthly Notices of Royal Astronomical Society* 173: 729–787.

[Heg84] Heggie D. C. (1984) Post-collapse evolution of a gaseous cluster model. *Monthly Notices of Royal Astronomical Society* 206: 179–195.

[Heg89] Heggie D. C. (1989) An n-body simulation of a gravothermal expansion. In *Dynamics of dense stellar systems*, pages 195–200. Cambridge University Press, Cambridge.

[Heg96] Heggie D. C. (1996) Statistics of small-n simulations. In Hut and Makino [HM96], pages 131–140.

[Hen71] Henon M. (1971) Monte-carlo models of star clusters. *Astrophysics and Space Science* 13: 284–299.

[Hen75] Henon M. (1975) Two recent developments concerning the monte carlo method. In *Dynamics of stellar systems*, pages 133–149. D. Reidel Publishing Co., Dordrecht.

[Her87] Hernquist L. (1987) Performance characteristics of tree codes. *The Astrophysical Journal Supplement Series* 64: 715–734.

[Her90] Hernquist L. (1990) Vectorization of tree traversals. *Journal of Computational Physics* 87: 137–147.

[HH95] Hozumi S. and Hernquist L. (1995) A comparison of two algorithms for simulating collisionless systems. *The Astrophysical Journal* 440: 60–68.

[HHM93] Hernquist L., Hut P., and Makino J. (1993) Discreteness noise versus force errors in n-body simulations. *The Astrophysical Journal Letters* 402: L85–+.

[Hic82] Hickson P. (1982) Systematic properties of compact groups of galaxies. *The Astrophysical Journal* 255: 382–391.

[Hil85] Hillis W. D. (1985) *The Connection Machine*. MIT Press, Cambridge, Massachusetts.

[HJ81] Hockney R. W. and Jesshope C. R. (1981) *Parallel Computers*. Adam Hilger, Ltd., Bristol.

[HK89] Hernquist L. and Katz N. (1989) Treesph – a unification of sph with the hierarchical tree method. *The Astrophysical Journal Supplement Series* 70: 419–446.

[HKW95] Hernquist L., Katz N., and Weinberg D. H. (1995) Physically detached 'compact groups'. *The Astrophysical Journal* 442: 57–60.

[HM86a] Heggie D. C. and Mathieu R. D. (1986) Standardised units and time scales. In Hut and McMillan [HM86b], pages 233–236.

[HM86b] Hut P. and McMillan S. (eds) (1986) *The Use of Supercomputers in Stellar Dynamics*, New York. Springer.

[HM96] Hut P. and Makino J. (eds) (1996) *Dynamical Evolution of Star Clusters*, Amsterdam. Kluwer.

[HMM88] Hut P., Makino J., and McMillan S. (1988) Modelling the evolution of globular star clusters. *Nature* 336: 31–35.

[HMM95] Hut P., Makino J., and McMillan S. (1995) Building a better leapfrog. *The Astrophysical Journal Letters* 443: L93–L96.

[HN78] Hachisu I. and Nakada Y. Sugimoto D. (1978) Gravothermal catastrophe of finite amplitude. *Progess of Theoretical Physics* 60: 393–402.

[HO92] Hernquist L. and Ostriker J. P. (1992) A self-consistent field method for galactic dynamics. *The Astrophysical Journal* 386: 375–397.

[Hos92] Hoshino T. (1992) The pax project. In Mendez R. (ed) *High Performance Computing: research and practice in Japan*, pages 239–256. John Weiley and Sons, Baffins Lane. This is a cross-referenced BOOK (collection) entry.

[HP90] Hennessy J. and Patterson D. (1990) *Computer Architecture: A Quantitative Approach*. Morgan Kaufmann, San Francisco.

[HR89] Heggie D. C. and Ramamani N. (1989) Evolution of star clusters after core collapse. *Monthly Notices of Royal Astronomical Society* 237: 757–783.

[HR92] Hut P. and Rees M. J. (1992) Constraints on massive black holes as dark matter candidates. *Monthly Notices of Royal Astronomical Society* 259: 27P–30P.

[HS78] Hachisu I. and Sugimoto D. (1978) Gravothermal catastrophe and negative specific head of self-gravitating systems. *Progess of Theoretical Physics* 60: 123–135.

[HSSC83] Hoogland A., Spaa J., Selman B., and Compagner A. (1983) A special-purpose processor for the monte-carlo simulation of ising spin systems. *Journal of Computational Physics* 51: 250–260.

[Ida90] Ida S. (1990) Stirring and dynamical friction rates of planetesimals in the solar gravitational field. *Icarus* 88: 129–145.

[IEMS91] Ito T., Ebisuzaki T., Makino J., and Sugimoto D. (1991) A special-purpose computer for gravitational many-body systems: Grape-2. *Publications of the Astronomical Society of Japan* 43: 547–555.

[IKM93] Ida S., Kokubo E., and Makino J. (1993) The origin of anisotropic velocity dispersion of particles in a disc potential. *Monthly Notices of Royal Astronomical Society* 263: 875–902.

[IM92a] Ida S. and Makino J. (1992) N-body simulation of gravitational interaction between planetesimals and a protoplanet. ii – dynamical friction. *Icarus* 98: 28–37.

[IM92b] Ida S. and Makino J. (1992) N-body simulation of gravitational interaction between planetesimals and a protoplanet. i – velocity distribution of planetesimals. *Icarus* 96: 107–120.

[IM93] Ida S. and Makino J. (1993) Scattering of planetesimals by a protoplanet: Slowing down of runaway growth. *Icarus* 106: 210–227.

[IMES90] Ito T., Makino J., Ebisuzaki T., and Sugimoto D. (1990) A special-purpose n-body machine grape-1. *Computer Physics Communications* 60: 187–194.

[IMF+93] Ito T., Makino J., Fukushige T., Ebisuzaki T., Okumura S. K., and Sugimoto D. (1993) A special-purpose computer for n-body simulations: Grape-2a. *Publications of the Astronomical Society of Japan* 45: 339–347.

[Jaf83] Jaffe W. (1983) A simple model for the distribution of light in spherical galaxies. *Monthly Notices of Royal Astronomical Society* 202: 995–999.

[Jer85] Jernigan J. G. (1985) Direct n-body simulations with a recursive center of mass reduction and regularization. In Goodman J. and Hut P. (eds) *Dynamics of Star Clusters; Proceedings of IAU Symposium 113*, pages 275–283. D. Reidel Publishing Co., Dordrecht.

[JP89] Jernigan J. G. and Porter D. H. (1989) A tree code with logarithmic reduction of force terms, hierarchical regularization of all variables, and explicit accuracy controls. *The Astrophysical Journal Supplement Series* 71: 871–893.

[Kan90] Kandrup H. E. (1990) Divergence of nearby trajectories for the gravitational n-body problem. *The Astrophysical Journal* 364: 420–425.

[KFTM97] Kawai A., Fukushige T., Taiji M., and Makino J. (1997) The pci interface for grape systems: Pci-hib. *Publications of the Astronomical Society of Japan* 49: 607–618.

[KI95] Kokubo E. and Ida S. (1995) Orbital evolution of protoplanets embedded in a swarm of planetesimals. *Icarus* 114: 247–257.

[KI96] Kokubo E. and Ida S. (1996) On runaway growth of planetesimals. *Icarus* 123: 180–191.

[KI97] Kokubo E. and Ida S. (1997) Oligarchic growth of protoplanets. submitted to Icarus.

[Kit86] Kittel C. (1986) *Introduction to Solid State Physics, 6th ed.* J. Wiley and Sons, Chichester.

[Kle97] Klessen R. (1997) Grapesph with fully periodic boundary conditions: Fragmentation of molecular clouds. *to appear in Monthly Notices of Royal Astronomical Society* .

[KMS93] Kandrup H. E., Mahon M. E., and Smith H. J. (1993) Energy and phase space mixing for self-gravitating systems of stars. *Astronomy and Astrophysics* 271: 440+.

[KMS94] Kandrup H. E., Mahon M. E., and Smith Haywood J. (1994) On the sensitivity of the n-body problem toward small changes in initial conditions. 4. *The Astrophysical Journal* 428: 458–465.

[Kor93] Koren I. (1993) *Computer Arithmetic Algorithms.* Prentice Hall, Englewood Cliffs, NJ.

[Kow85] Kowalik J. (ed) (1985) *Parallel MIMD Computation: The HEP Supercomputer and Its Applications.* MIT Press, Boston.

[KR71] Kingsbury N. G. and Rayner P. J. W. (1971) Digital filtering using logarithmic arithmetic. *Electronics Letters* 7: 56–58.

[Kro74] Krogh F. T. (1974) Changing stepsize in the integration of differential equations using modified divided differences. In Bettis D. G. (ed) *Proceedings of the Conference on the Numerical Solution of Ordinary Differential Equations*, number 362 in Lecture Notes in Mathematics, pages 22–71. Springer-Verllag, New York.

[KS65] Kustaanheimo P. and Stiefel E. (1965) Perturbation theory of kepler motion based on spinor regularization. *J. Reine. Angew. Math.* 218: 204–219.

[KYM97] Kokubo E., Yoshinaga K., and Makino J. (1997) On a time-symmetric hermite integrator for planetary n-body simulation. *submitted to Monthly Notices of Royal Astronomical Society* .

[LAB+95] Lauer T. R., Ajhar E. A., Byun Y. I., Dressler A., Faber S. M., Grillmair C., Kormendy J., Richstone D., and Tremaine S. (1995) The centers of early-type galaxies with hst.i.an observational survey. *The Astronomical Journal* 110: 2622+.

[Lau85] Lauer T. R. (1985) The cores of elliptical galaxies. *The Astrophysical Journal* 292: 104–121.

[LD94] Levison H. F. and Duncan M. J. (1994) The long-term dynamical behavior of short-period comets. *Icarus* 108: 18–36.

[LE80] Lynden-Bell D. and Eggleton P. P. (1980) On the consequences of the gravothermal catastrophe. *Monthly Notices of Royal Astronomical Society* 191: 483–498.

[Lis93] Lissauer J. J. (1993) Planet formation. *Annual Review of Astronomy and Astrophysics* 31: 129–174.

[LO85] Lacey C. G. and Ostriker J. P. (1985) Massive black holes in galactic halos? *The Astrophysical Journal* 299: 633–652.

[LS77] Lightman A. P. and Shapiro S. L. (1977) The distribution and consumption rate of stars around a massive, collapsed object. *The Astrophysical Journal* 211: 244–262.

[LSH71] Lyman Spitzer J. and Hart M. H. (1971) Random gravitational encounters and the evolution of spherical systems. i. method. *The Astrophysical Journal* 164: 399–409.

[LW68] Lynden-Bell D. and Wood R. (1968) The gravo-thermal catastrophe in isothermal spheres and the onset of red-giant structure for stellar systems. *Monthly Notices of Royal Astronomical Society* 138: 495–525.

[LW95] Leonski D. E. and Wever W.-D. (1995) *Scalable Shared-Memory Multi-processing*. Morgan Kaufmann, San Francisco.

[Lyn67] Lynden-Bell D. (1967) Statistical mechanics of violent relaxation in stellar systems. *Monthly Notices of Royal Astronomical Society* 136: 101–121.

[MA92] Makino J. and Aarseth S. J. (1992) On a hermite integrator with ahmadcohen scheme for gravitational many-body problems. *Publications of the Astronomical Society of Japan* 44: 141–151.

[MA93a] McMillan S. L. W. and Aarseth S. J. (1993) An o(n log n) integration scheme for collisional stellar systems. *The Astrophysical Journal* 414: 200–212.

[MA93b] Mikkola S. and Aarseth S. J. (1993) An implementation of n-body chain regularization. *Celestial Mechanics and Dynamical Astronomy* 57: 439–459.

[Mak89] Makino J. (1989) Gravothermal oscillations in n-body systems. In *Dynamics of dense stellar systems*, pages 201–206. Cambridge University Press, Cambridge.

[Mak90] Makino J. (1990) Vectorization of a treecode. *Journal of Computational Physics* 87: 148–160.

[Mak91a] Makino J. (1991) A modified aarseth code for grape and vector processors. *Publications of the Astronomical Society of Japan* 43: 859–876.

[Mak91b] Makino J. (1991) Optimal order and time-step criterion for aarseth-type n-body integrators. *The Astrophysical Journal* 369: 200–212.

[Mak91c] Makino J. (1991) Treecode with a special-purpose processor. *Publications of the Astronomical Society of Japan* 43: 621–638.

[Mak96] Makino J. (1996) Gravothermal oscillations. In Hut and Makino [HM96], pages 151–160.

[Mak97] Makino J. (1997) Merging of galaxies with central black holes. ii. evolution of the black hole binary and the structure of the core. *The Astrophysical Journal* 478: 58+.

[Man87] Mann P. J. (1987) Finite difference methods for the classical particle-particle gravitational n-body problem. *Computer Physics Communications* 47: 213–228.

[MAS90] Makino J., Akiyama K., and Sugimoto D. (1990) On the apparent universality of the r exp 1/4 law for brightness distribution in galaxies. *Publications of the Astronomical Society of Japan* 42: 205–215.

[Mas95] Masuda N. (1995) Error analysis of special-purpose computer for sph (in japanese). Master's project, University of Tokyo, Department of Earth Science and Astronomy.

[McC95] McCulpin J. D. (1995) Memory bandwidth and machine balance in current high performance computers. *Newsletter of IEEE Computer Architecture Technical Committee* 1: 19–25.

[McM86] McMillan S. L. W. (1986) The vectorization of small-n integrators. In Hut and McMillan [HM86b], pages 156–161.

[ME94] Makino J. and Ebisuzaki T. (1994) Triple black holes in the cores of galaxies. *The Astrophysical Journal* 436: 607–610.

[ME96] Makino J. and Ebisuzaki T. (1996) Merging of galaxies with central black holes. i. hierarchical mergings of equal-mass galaxies. *The Astrophysical Journal* 465: 527–533.

[Mer96] Merritt D. (1996) Optimal smoothing for n-body codes. *The Astronomical Journal* 111: 2462+.

[MF93] Makino J. and Funato Y. (1993) The grape software system. *Publications of the Astronomical Society of Japan* 45: 279–288.

[MFOE93] Makino J., Fukushige T., Okumura S. K., and Ebisuzaki T. (1993)

The evolution of massive black-hole binaries in merging galaxies. i – evolution of a binary in a spherical galaxy. *Publications of the Astronomical Society of Japan* 45: 303–310.

[MH88] Makino J. and Hut P. (1988) Performance analysis of direct n-body calculations. *The Astrophysical Journal Supplement Series* 68: 833–856.

[MH89] Makino J. and Hut P. (1989) Gravitational *n*-body algorithms: A comparison between supercomputers and a highly parallel computer. *Computer Physics Reports* 9: 199–246.

[MH90] Makino J. and Hut P. (1990) Bottlenecks in simulations of dense stellar systems. *The Astrophysical Journal* 365: 208–218.

[MH97] Makino J. and Hut P. (1997) Merger rate of equal-mass spherical galaxies. *The Astrophysical Journal* 481: 83+.

[MHM90] McMillan S., Hut P., and Makino J. (1990) Star cluster evolution with primordial binaries. i – a comparative study. *The Astrophysical Journal* 362: 522–537.

[MHM91] McMillan S., Hut P., and Makino J. (1991) Star cluster evolution with primordial binaries. ii – detailed analysis. *The Astrophysical Journal* 372: 111–124.

[MIE90] Makino J., Ito T., and Ebisuzaki T. (1990) Error analysis of the grape-1 special-purpose n-body machine. *Publications of the Astronomical Society of Japan* 42: 717–736.

[Mil64] Miller R. H. (1964) Irreversibility in small stellar dynamical systems. *The Astrophysical Journal* 140: 250–256.

[MKT93] Makino J., Kokubo E., and Taiji M. (1993) Harp: A special-purpose computer forn-body problem. *Publications of the Astronomical Society of Japan* 45: 349–360.

[MTES97] Makino J., Taiji M., Ebisuzaki T., and Sugimoto D. (1997) Grape-4: A massively parallel special-purpose computer for collisional n-body simulations. *The Astrophysical Journal* 480: 432+.

[MTTS94] Makino J., Taiji M., T.Ebisuzaki, and Sugimoto D. (1994) Grape-4: A one-tflops special-purpose computer for astrophysical *n*-body problems. In *Proceedings Supercomputing '94*, pages 429–438. IEEE, Los Alamitos.

[MV92] Mikkola S. and Valtonen M. J. (1992) Evolution of binaries in the field of light particles and the problem of two black holes. *Monthly Notices of Royal Astronomical Society* 259: 115–120.

[MYTN97] Mori M., Yoshii Y., Tsujimoto T., and Nomoto K. (1997) The evolution of dwarf galaxies with star formation in an outward-propagating supershell. *The Astrophysical Journal Letters* 478: L21–+.

[NFW96] Navarro J. F., Frenk C. S., and White S. D. M. (1996) The structure of cold dark matter halos. *The Astrophysical Journal* 462: 563+.

[NS97] Navarro J. F. and Steinmetz M. (1997) The effects of a photoionizing ultraviolet background on the formation of disk galaxies. *The Astrophysical Journal* 478: 13+.

[NW93] Navarro J. E. and White S. D. M. (1993) Simulations of dissipative galaxy formation in hierarchically clustering universes – 1. tests of the code. *Monthly Notices of Royal Astronomical Society* 265: 271–378.

[OEM91] Okumura S. K., Ebisuzaki T., and Makino J. (1991) Kinematic structures of merger remnants. *Publications of the Astronomical Society of Japan* 43: 781–793.

[OM85] Ogielski A. T. and Morgenstern I. (1985) Critical behavior of three-dimentional ising spin-glass model. *Phys. Rev. Lett.* 54: 928–931.

[OP73] Ostriker J. and Peebles P. J. E. (1973) A numerical study of the stability of flattened galaxies: Or, can cold galaxies survive? *The Astrophysical Journal* 186: 467–480.

[PAC+97] Patterson D., Anderson T., Cardwell N., Fromm R., Keeton K., Kozyrakis C., Thomas R., and Yelik K. (1997) A case for intelligent ram. *IEEE Micro* 17(2): 34–43.

[PG96] Pfalzner S. and Gibbon P. (1996) *Many-Body Tree Methods in Physics*. Cambridge University Press, Cambridge.

[PRT83] Pearson R. B., Richardson J. L., and Toussaint D. (1983) A fast processor for monte-carlo simulation. *Journal of Computational Physics* 51: 241–249.

[PTVF92] Press W. H., Teukolski S. A., Vetterling W. T., and Flannery B. P. (1992) *NUMERICAL RECIPES in C*. Cambridge University Press, Cambridge, second edition.

[Qui96] Quinlan G. D. (1996) The dynamical evolution of massive black hole binaries i. hardening in a fixed stellar background. *New Astronomy* 1: 35–56.

[Ric93] Richardson D. C. (1993) A new tree code method for simulation of planetesimal dynamics. *Monthly Notices of Royal Astronomical Society* 261: 396–414.

[Ric94] Richardson D. C. (1994) Tree code simulations of planetary rings. *Monthly Notices of Royal Astronomical Society* 269: 493+.

[RT96] Rauch K. P. and Tremaine S. (1996) Resonant relaxation in stellar systems. *New Astronomy* 1: 149–170.

[SA75] Swartzlander E. E. and Alexopoulos A. E. (1975) The sign/logarithm number system. *IEEE Transactions on Computer* C-24: 1238–1242.

[SA96] Spurzem R. and Aarseth S. J. (1996) Direct collisional simulation of 10000 particles past core collapse. *Monthly Notices of Royal Astronomical Society* 282: 19–39.

[SB83] Sugimoto D. and Bettwieser E. (1983) Post-collapse evolution of globular clusters. *Monthly Notices of Royal Astronomical Society* 204: 19P–22P.

[Sel96] Seljak U. (1996) Gravitational lensing effect on cosmic microwave background anisotropies: A power spectrum approach. *The Astrophysical Journal* 463: 1+.

[SHQ95] Sigurdsson S., Hernquist L., and Quinlan G. D. (1995) Models of galaxies with central black holes: Simulation methods. *The Astrophysical Journal* 446: 75+.

[SIA94] SIA (1994) The national technology loadmap for semiconductors. Technical report, Semiconductor Industry Association, San Jose, CA.

[SK95] Sosin C. and King I. R. (1995) Hst observations of the core of the globular cluster ngc 6624. *The Astronomical Journal* 109: 639–649.

[SM89] Sugimoto D. and Makino J. (1989) Synchronization instability and merging of binary globular clusters. *Publications of the Astronomical Society of Japan* 41: 1117–1144.

[SM94] Steinmetz M. and Mueller E. (1994) The formation of disk galaxies in a cosmological context: Populations, metallicities and metallicity gradients. *Astronomy and Astrophysics* 281: L97–L100.

[SM95] Steinmetz M. and Muller E. (1995) The formation of disc galaxies in a cosmological context: structure and kinematics. *Monthly Notices of Royal Astronomical Society* 276: 549–562.

[SMS95] Sterling T., Messina P., and Smith P. H. (1995) *Enabling Technologies for Petaflops Computing*. MIT Press, Cambridge, MA.

[SMVO96] Shapiro P. R., Martel H., Villumsen J. V., and Owen J. M. (1996)

Adaptive smoothed particle hydrodynamics, with application to cosmology: Methodology. *The Astrophysical Journal Supplement Series* 103: 269+.

[Spi96] Spitzer Lyman J. (1996) *Dynamical Evolution of Globular Clusters.* Princeton University Press, Princeton, New Jersey.

[SSC94] Sanz-Serna J. M. and Calvo M. P. (1994) *Numerical Hamiltonian Problems.* Chapman and Hall, London.

[ST92] Saha P. and Tremaine S. (1992) Symplectic integrators for solar system dynamics. *The Astronomical Journal* 104: 1633–1640.

[ST94] Saha P. and Tremaine S. (1994) Long-term planetary integration with individual time steps. *The Astronomical Journal* 108: 1962–1969.

[Ste96] Steinmetz M. (1996) Grapesph: cosmological smoothed particle hydrodynamics simulations with the special-purpose hardware grape. *Monthly Notices of Royal Astronomical Society* 278: 1005–1017.

[SW88a] Stewart G. R. and Wetherill G. W. (1988) Evolution of planetesimal velocities. *Icarus* 74: 542–553.

[SW88b] Sussman G. J. and Wisdom J. (1988) Numerical evidence that the motion of pluto is chaotic. *Science* 241: 433–437.

[SW94] Salmon J. K. and Warren M. S. (1994) Skeletons from the treecode closet. *Journal of Computational Physics* 111: 136–155.

[Swe93] Sweatman W. L. (1993) A study of lagrangian radii oscillations and core-wandering using n-body simulations. *Monthly Notices of Royal Astronomical Society* 261: 497–512.

[Tan87] Tanekusa J. (1987) Statistical theory of violent relaxation. *Publications of the Astronomical Society of Japan* 39: 425–436.

[Ter88] Terrano A. E. (1988) The qcd machine. In Alder B. J. (ed) *Special Purpose Computers*, pages 41–67. Academic Press, San Diego.

[Teu95] Teuben P. (1995) The stellar dynamics toolbox nemo. In Shaw R., Payne H., and Hayes J. (eds) *Astronomical Data Analysis Software and Systems IV*, volume 77 of *ASP Conference Series*, pages 398+.

[TIN96] Tanaka H., Inaba S., and Nakazawa K. (1996) Steady-state size distribution for the self-similar collision cascade. *Icarus* 123: 450–455.

[TIS88] Taiji M., Ito N., and Suzuki M. (1988) Special purpose computer system ddoe ising models. *Reviews of Scientific Instruments* 59: 2379–2384.

[Too77] Toomre A. (1977) ... In Tinsley B. M. and Larson R. B. (eds) *Evolution of Galaxies and Stellar Populations*, page 401. Yale Observatory, New Heaven.

[TW96] Taniguchi Y. and Wada K. (1996) The nuclear starburst driven by a supermassive black hole binary. *The Astrophysical Journal* 469: 581+.

[UFM+93] Umemura M., Fukushige T., Makino J., Ebisuzaki T., Sugimoto D., Turner E. L., and Loeb A. (1993) Smoothed particle hydrodynamics with grape-1a. *Publications of the Astronomical Society of Japan* 45: 311–320.

[Van82] Van Albada T. S. (1982) Dissipationless galaxy formation and the r to the 1/4-power law. *Monthly Notices of Royal Astronomical Society* 201: 939–955.

[Wei94] Weinberg M. D. (1994) Adiabatic invariants in stellar dynamics, 3: Application to globular cluster evolution. *The Astronomical Journal* 108: 1414–1420.

[WH91] Wisdom J. and Holman M. (1991) Symplectic maps for the n-body problem. *The Astronomical Journal* 102: 1528–1538.

[Whi78] White S. D. M. (1978) Simulations of merging galaxies. *Monthly Notices of Royal Astronomical Society* 184: 185–203.

[Whi79] White S. D. M. (1979) Can mergers make slowly rotating elliptical galaxies. *The Astrophysical Journal Letters* 229: L9–L13.

[WS89] Wetherill G. W. and Stewart G. R. (1989) Accumulation of a swarm of small planetesimals. *Icarus* 77: 330–357.

[WT88] Wisdom J. and Tremaine S. (1988) Local simulations of planetary rings. *The Astronomical Journal* 95: 925–940.

[XO94] Xu G. and Ostriker J. P. (1994) Dynamics of massive black holes as a possible candidate of galactic dark matter. *The Astrophysical Journal* 437: 184–193.

[YGBS94] Yanny B., Guhathakurta P., Bahcall J. N., and Schneider D. P. (1994) Globular cluster photometry with the hubble space telescope. 2: U, v, and i measurements of m15. *The Astronomical Journal* 107: 1745–1763.

[YOT⁺96] Yokono Y., Ogasawara R., Takeuchi T., Inutsuka S., Miyama S. M., and Chikada Y. (1996) Development of special-purpose computer for cosmic hydrodynamics by sph. In Tomisaka K. (ed) *Numerical Astrophysics Using Supercomputers*. National Astronomical Observatory, Japan.

[You80] Young P. (1980) Numerical models of star clusters with a central black hole. i – adiabatic models. *The Astrophysical Journal* 242: 1232–1237.

Index